PROGRAMMING MOBILE DEVICES

PROGRAMMING MOBILE DEVICES

AN INTRODUCTION FOR PRACTITIONERS

Tommi Mikkonen
Tampere University of Technology, Finland

John Wiley & Sons, Ltd

Other Wiley Editorial Offices

John Wiley & Sons Inc., 111 River Street, Hoboken, NJ 07030, USA

Jossey-Bass, 989 Market Street, San Francisco, CA 94103-1741, USA

Wiley-VCH Verlag GmbH, Boschstr. 12, D-69469 Weinheim, Germany

John Wiley & Sons Australia Ltd, 42 McDougall Street, Milton, Queensland 4064, Australia

John Wiley & Sons (Asia) Pte Ltd, 2 Clementi Loop #02-01, Jin Xing Distripark, Singapore 129809

John Wiley & Sons Canada Ltd, 6045 Freemont Blvd, Mississauga, Ontario, L5R 4J3, Canada

Wiley also publishes its books in a variety of electronic formats. Some content that appears
in print may not be available in electronic books.

Anniversary Logo Design: Richard J. Pacifico

Library of Congress Cataloging-in-Publication Data:

Mikkonen, Tommi.
   Programming mobile devices : an introduction for practitioners /
Tommi Mikkonen.
         p. cm.
   Includes bibliographical references and index.
   ISBN 978-0-470-05738-4 (cloth : alk. paper)
   1. Mobile computing. 2. Wireless communication systems. I. Title.
   QA76.59M54 2007
   004.165 – dc22
                                        2006036202

British Library Cataloguing in Publication Data

A catalogue record for this book is available from the British Library

ISBN 978-0-470-05738-4 (Hb)

Typeset in 10/12 Times by Laserwords Private Limited, Chennai, India
Printed and bound in Great Britain by Antony Rowe Ltd, Chippenham, Wiltshire
This book is printed on acid-free paper responsibly manufactured from sustainable forestry
in which at least two trees are planted for each one used for paper production.

Contents

Foreword by Jan Bosch

In 1968, a NATO-organized conference was held during which several terms central to our field were introduced, including software component, software architecture and software engineering. This conference can be viewed as a milestone in turning the programming of software systems from a craft to a true engineering discipline. Although we have not yet reached a level of proficiency that is on par with the older engineering disciplines, enormous progress has been made in the last four decades, allowing us to build software systems, even complete ecosystems, that go far beyond the dreams of the first software engineers. This book marks a similar milestone where the construction of mobile systems is moving from a craft to an engineering discipline.

A key characteristic of any engineering discipline is that its professionals have the ability to build and evolve systems that a layman would either not be capable of or can only construct at a productivity level that is one or several orders of magnitude lower. The software engineering professional combines a deep insight into the fundamental principles underlying the discipline, such as modularity, composability, architecture, quality and user experience, as well as a detailed knowledge of the strengths and limitations of the mechanical and hardware systems for which the software is developed and the tools used to develop software.

Since a few months ago, my oldest son, age 9, is the proud owner of a Series 60 Nokia mobile phone. Interested in mathematics and computer games, his immediate questions assorted to the ability to download software, arguably not productivity applications but rather games, as well as the possibility to develop software for the phone himself. Although our joint programming efforts have been limited to Python, rather than Java and C++ as discussed in this book, reflecting on the discussions with my son reinforced my realization to what extent and at what speed computing is moving to the edge of the network, specifically to mobile devices. It is of course a cliché, but that is because it's true!

The trends of convergence and mobility have a profound impact on society. The typical adoption pattern for new use cases, e.g. making a phone call, taking a picture, listening to music or reading and sending email, consists of three phases. At first, some use case is simply not feasible when using an integrated mobile device. During the second stage, the use case becomes possible using a mobile device, but it is the second choice for the user, because the experience of use and the associated cost are significantly behind the stationary or single purpose devices. During the third stage, the specific use case matures and becomes the preferred choice. An illustrative example is the basic phone call. Not available at all in most of the western world until the early 1990s, mobile phones were initially used where access to fixed phones was lacking and the importance of communication warranted no delay. Today we see ubiquitous use of mobile phones and most would prefer using a mobile phone even while standing next to a fixed phone. There are numerous use cases going through the same three-staged adoption process including taking and watching pictures as well as video, sending and receiving messages, including SMS, email and instant messaging, gaming, access to enterprise applications and business processes, surfing and searching the internet, watching television, participating in online communities and, in general, interacting with the increasingly fusing digital and physical worlds.

Summarizing, we see convergence, i.e. the integration of use cases earlier reserved for single purpose devices, as a very strong trend set to continue. Second, mobility, i.e. transferring stationary and nomadic use cases to true mobile contexts, is freeing individuals from the confines of specific locations, e.g. an office desk, and specific contexts, e.g. sitting in an airport lounge with a laptop computer in, well, one's lap. The ability to perform use cases when and wherever is not just good for productivity in enterprise contexts, but also hugely satisfying from a personal perspective. The third trend is that a mobile device is personal and defining the identity of its user to an extent that goes far beyond desktop or laptop computers. Consequently, we see mobile devices differ in size, form factor, input and output devices, available built-in hardware, e.g. GPS, wireless LAN and Bluetooth, external accessories, interaction with external devices, etc. to extent far beyond traditional computing.

In the discussion so far, I have tried to build a case for my conviction that software engineering has reached a next major milestone or perhaps even paradigm shift, to use Thomas Kuhn's terminology, in the shift to mobile computing. Even though there already is significant attention to the topic, we are only at the very beginning of a major transformation in the information technology industry. This transformation will require unprecedented degrees of adaptability, configurability and composability of software. Already today, we can see that software developed for mobile devices requires many versions in order to handle the variations between different mobile devices. Second, as mobile software is used in many different contexts, software services and applications lack the ability to intelligently adjust their behaviour to the current context. The simple example of a too loud ring tone in a meeting or a too soft one while having a drink in a noisy bar illustrates this. Third, the user, being mobile, is constantly in the presence of a constantly changing set of stationary, nomadic or mobile devices that can be communicated with and used by the user's mobile device to improve its user's efficiency and ability to perform tasks. Finally, mobile devices are personal and consequently require a high degree of personalization that also affects the software on the device.

The key activity central to the transformation to mobile computing is the programming of applications and services on mobile devices. One can take three perspectives to mobile software. From the first perspective, nothing changes – software is software and the same programming languages and basic principles apply. From the second perspective, one could take the position that software for mobile devices is a step back from desktop software in that the software engineer needs to consider, among others, resource constraints, user interaction, memory management and security solutions in a way that is similar to desktop software about a decade or more ago. The third perspective that one can take, and that I feel is the more appropriate one, is that mobile software adds a unique and novel dimension to programming that has not been present before and that requires new programming practices and approaches that, being early in this transformation, we are currently only starting to explore and experiment with.

The book that you are currently holding marks this milestone in the evolution of software engineering in an exquisite manner. The author, Tommi Mikkonen, has managed to strike the delicate balance between defining and discussing the principles that are fundamental to mobile software on the one hand and on the other hand discuss the concrete details of programming languages, specifically mobile Java and Symbian C++, that can be used for programming Series 60 mobile devices. Since the Series 60 platform is the absolute market leader in open, mid- to high-end mobile devices, this book is a must-read for anyone interested in programming these devices. The second advantage of the approach taken by the author is that the book manages to describe the details of specific releases of programming environments without making the reader dependable on the specific version. The principles help programmers to easily evolve to subsequent versions, which appear at a very high rate, i.e. multiple times per year.

I warmly recommend this book to anyone interested in programming mobile devices or interested in the state of the art and practice in this area. We are at a the verge of a major transformation in the information technology industry towards mobile computing and this book represents and outlines this future in clear and detailed fashion that will leave the reader with a solid understanding of programming for mobile devices.

Jan Bosch
Head of Software and Application Technologies Laboratory
Nokia Research Center
Helsinki, Finland.
October 2006

Foreword by Antero Taivalsaari

I've known Tommi for several years, both as a colleague and as a friend. Ever since first meeting him, Tommi has been passionate about mobile devices and mobile software development, not only as a professor and an academic researcher, but also as an enthusiastic mobile software developer himself.

Tommi has pioneered the teaching of mobile software development in Finland. He arranged the first university-level courses on mobile software development in Finland back in 2001, and in the past years he has instructed over a thousand students to become proficient in this exciting and rapidly evolving field. Unfortunately, the extensive lecture material that Tommi has prepared for his mobile software development courses has been available only in Finnish so far.

In this book, Tommi makes his expertise in mobile software development available also to English-speaking software developers and students. The book presents a comprehensive summary of all the central areas in mobile software development, ranging from fundamental topics such as memory and resource management to application design, networking, concurrency and security.

Rather than focusing on specific technologies, devices or operating systems, this book takes a different approach and presents a summary of the component areas and issues that are common to all the mobile software platforms. This should result in a more "timeless" book that should stand the test of time well, unlike so many other books that are outdated already by the time they come out of press.

Knowing Tommi's hectic schedule, this book represents a massive undertaking from Tommi's part. I am both impressed and envious about his ability to write such a book, while working on so many other projects simultaneously. I am confident that this book, for its part, will make the exciting area of mobile software development more approachable to a new generation of students and software developers worldwide.

Dr. Antero Taivalsaari
The original designer of the Java™ Platform, Micro Edition (Java ME)
Sun Microsystems Laboratories

Preface

The two latest decades have seen the introduction of more and more hand-held gadgets being used for communication, as personal digital assistants, and simply for fun. Personal digital assistants and mobile phones, followed by other types of devices, such as MP3 players, wrist-watches and the like, have been adopted for wide use in a relatively short period of time. During the time frame of these decades, these devices have encountered a major change in their design; many devices were first fabricated predominantly with hardware, and they served a single purpose. More recently, as the computing power in them has increased to the level of state-of-the-art desktops only some years ago, the devices have become a programming environment that has emerged as a new domain of software development, to the extent that one can even add new software developed by a designer independent of the device manufacturer. Moreover, properly documented programming infrastructure has been introduced to allow one to introduce programming facilities to a proprietary system without risking the features of the original device.

The outcomes of improved facilities included in modern mobile devices are many. One can obviously introduce personalized features in devices, and thus create a system that is best suited for some particular use. Moreover, also mass customization becomes possible, as copying software once it has been completed is virtually free. In addition to personal use, also commercial use by enterprises becomes more tempting, as it is possible to create systems that extend from enterprises' servers to all employees anywhere and any time. Starting with email, already now a number of enterprises are allowing more and more mobile personnel who can interact with company intranet and computing systems disregarding the restrictions of time and place. Moreover, the number of devices that forms the potential market for new applications is huge, and being able to attract a fraction of it will lead to success.

For a company implementing such devices, a major challenge is introduced in the form of transforming companies that develop hardware-based systems that included small portions of software to predominantly software companies. Firstly, software in mobile devices should be robust and reliable to allow users to depend on it. Secondly, at the same time, software should be generic so that it can be used in as many devices as possible to allow the largest user base possible to reduce the need for new development. Furthermore, changes that are evident due to evolving hardware – more memory and processing power is regularly introduced in newer devices, together with more sophisticated hardware features like a camera – should not raise any compatibility issues but the same code should still run. Finally, the sheer size of needed software has grown from a small portion to tens of megabytes in some devices. Managing this amount of software in a practical yet cost-effective fashion is a grand challenge.

Although programming is really fundamentally the same, be a program targeted to a mobile device environment or to a desktop, resources available in the latter can be considered a lot more forgiving in the sense that a smallish programming error resulting in garbaging some memory every now and then will probably never cause a failure. In contrast, a smallish error in a program targeted for a mobile device in a part that handles memory use can cause devastating effects due to the restricted resources of the device. Therefore the quality of a design, including also non-functional properties, is even more important in the mobile setting. As a result, designs will unfortunately be harder to compose. Moreover, although the users of mobile devices are often adopting usage patterns of embedded devices where they expect the device to react immediately, when using a programmable mobile device, delays in execution will inevitably occur as old applications are being shut down and new ones started. Luckily, when aiming at the development of a single application, ideally for a single device that one owns, the task is often not overly complex, but can be performed with minor effort. Moreover, people tend to be more forgiving with regard to features specialized by and for themselves.

Another way to look at programming of mobile devices is that in some ways it is about putting together some pieces of embedded systems development. In an embedded environment, one often has to compose programs in a memory-aware fashion, and take into account that the system can be active for a virtually unlimited time. Moreover, in such settings, limited performance of hardware is commonly assumed as a starting point. In the workstation environment, on the other hand, long-living application development platforms are a commodity we have become used to. Furthermore, applications are designed with modifications and future additions in mind. An additional factor is the development time, which has led us to update workstation software on an almost daily basis; in fact, this can be automated. Not surprisingly, a similar need for constant management exists when considering the use of mobile devices in a professional context. However, there is an important difference. In the desktop setting, applications used by corporations can be managed by an IT department that determines what kinds of applications can be allowed, as well as the settings that can be used when running the application. For a desktop environment, several systems are available for managing the application setting. However, such systems have become available for managing the applications in mobile devices only recently. Furthermore, even when relying on device management, a user may still have full access to everything in the device, and even if the corporation were able to install and configure an application, the user can remove and reconfigure the application. Overall, this can in fact be considered as a major obstacle for using mobile applications in the corporate context.

Increasing dependability requirements of communication are also applicable to the mobile setting. It is obvious that mobile devices can already now run applications that are extensions of existing systems benefiting from mobility. In contrast, however, another perspective to using mobile devices as application environment is to implement new, small, yet innovative systems that run only (or predominantly) in mobile devices. Currently, mobile games, which already now have established a foothold as a major line of business, are probably the best representative of this approach. Even with the current devices on the market, the number of potential users of such pieces of software is high, and therefore, the price of one download can be relatively low. The lower price of a sold application, say $10 per download, can be compensated with the larger number of downloads.

Based on the above, programming of mobile devices differs from other domains in its characteristics. This book is intended as a textbook on the principles of designing software for mobile devices. It is targeted at programmers who have experience in application development, but who have not worked with mobile devices before. The book is based on experiences in working in the mobile devices industry as well as experiences in teaching programming of mobile devices at Tampere University of Technology for several years. Unlike many other textbooks on programming mobile devices, the book is not intended to be used as a guide for immediately writing programs or creating applications for a certain mobile platform. Rather, the goal is to introduce the main ideas and restrictions that are applicable in any mobile environment, thus enabling more generic use. Moreover, the presentation does not become invalidated when a new version of a platform comes out that introduces different facilities for some parts of the system. Still, existing platforms are used as examples on how to compose actual mobile software and on how the discussed principles are visible in practical implementations. The discussion is structured as an introduction to the mobile devices infrastructure in general, memory and its use, applications and their development, modularity based on dynamically linked libraries, concurrency, resources and their management, networking, and security. Each category will be addressed in a chapter of its own.

Finally, let us consider a fundamental question: is there room for special practice of mobile devices programming, or will it be similar to programming a desktop or laptop computer? Even now, selecting a very computer-like mobile device for some particular special-purpose application is an option. Furthermore, with such devices, restrictions related to scarce resources can be relaxed, as it is possible to purchase more memory in order to make the devices optimal for a particular application. In addition, also programming environments that are used in workstations have been proposed for the mobile environment, including Python and Visual Basic, for instance. A further option would be to use a system where an operating system resembling those of PCs was available, like a restricted form of Windows and Linux, where application development can be more familiar. Moreover, using for example JavaScript inside a browser can also be considered yet another way to compose programs for mobile devices in a fashion that does not differ much from workstation programming. However, when aiming at the development of software that can be used in practice in a maximum number of devices, it is more likely that the restrictions remain, as the hardware forms a considerable cost factor and not all phones include broadband connectivity, but only something more modest. Still they require software to operate. Therefore, despite the advances in high-end phones and their connectivity features, as long as there is room for cost-effectively manufactured phones that are restricted in their performance, programming them in a sophisticated fashion requires special skills.

Acknowledgments

For being able to compose a book on programming mobile devices, I am grateful to a number of people in my professional life. I especially wish to express my sincere thanks to Dir. Terho Niemi for putting me in charge of mobile device software architecture, which has initiated all this work; Prof. Ilkka Haikala for convincing me to apply for a professorship on mobile device programming; colleagues who have enabled many inspiring discussions; a number of master and doctoral students composing their theses on mobile systems programming; all the staff who have participated in the class of mobile programming in one form or another; and all the students who have taken the class and thus contributed to improving the material.

I also wish to thank a number of people who have reviewed different versions and parts of this book, including in particular Kari Systä, Tino Pyssysalo, Antti Juustila, and Reino Kurki-Suonio. Thanks to Symbian reviewers (Richard Harrison, Rick Martin, Rahul Singh, Kostyantyn Lutsenko, Warren Day, Ioannis Dourus, Attila Vamos, Jonathan Yu, Leela Prasanna, Krishna Vasudevan, Kamal Singhania, Kavita Khatawate, Amit Shivnani, Kajal Ahuja). I also wish to thank Laserwords for their help during the production process.

Finally, I also have a personal life. To this end, I wish to express my sincere apologies to my family. Hopefully daddy will be around more often in the future.

Tommi Mikkonen
Tampere, Finland.

1

Introduction

1.1 Motivation

The development of mobile software has often been addressed in a fashion that focuses on using some particular technologies. While this type of approach can be easily justified for the introduction of a mobile platform that is to be used as the basis of an implementation, long-term issues are harder to embed into such an introduction. Furthermore, as the number of mobile platforms has been increasing, it is becoming an option to aim at discussing the differences between workstation and embedded software and software that runs in mobile devices at a general rather than at an implementation-specific level. We believe that this leads to a longer lasting approach, which will not be outdated when a new version of some particular mobile platform is introduced, since the basic patterns and philosophy of a design are likely to remain the same even if the platform version changes.

Principally, the design of software that runs in a mobile device requires that developers combine the rules of thumb applicable in the embedded environment – memory awareness, turned on for an unlimited time, limited performance and resources in general, and security in the sense that the device should never malfunction to produce unanticipated costs or reveal confidential information even if the user behaves in an unanticipated fashion – with features that are needed in the workstation environment – modifiability and adaptability, run-time extensions, and rapid application development. For this combination, the designer must master both hardware-aware and application-level software, as well as the main principles that guide their design. In order to compose designs where all these requirements are satisfied, the designer is bound to use abstraction, which is the most powerful weapon for dealing with complexity.

1.1.1 Leaking Abstractions

Due to being such a powerful weapon for attacking software development, abstraction is also one of the most commonly used facilities in programming. Systems we

Programming Mobile Devices: An Introduction for Practitioners   Tommi Mikkonen
© 2007 John Wiley & Sons, Ltd

commonly use are full of abstractions, such as menus, databases, or file systems, to name a few. Moreover, we are good at managing abstractions we are familiar with, and know how they should be used. Therefore, the skill of programming in a certain environment implies that one recognizes the basic abstractions applied in the environment, and knows how the abstractions are intended to be used.

Unfortunately, abstractions are not problem-free. In particular, problems are imminent when we face a new application domain or environment, such as mobile devices. Commonly used abstractions of programming may no longer be solid but they can start to leak. We will study this phenomenon in more detail in the following.

In principle, as argued by several authors, including Gannon et al. (1981) and Gabriel (1989) for instance, the user of an abstraction can overlook the details of the underlying implementation. In practice, however, when composing programs, details of the underlying hardware and underlying infrastructure software used as the implementation technique of a certain abstraction sometimes become visible to the software developer or even to the user of the system. We will call this leaking abstraction.[1] However, without knowing the implementation, it is difficult to understand what happens when the program is executed and a sudden downgrade of performance occurs, for instance. This makes the design more difficult, as revealing the implementation can take alarming forms.

As a sample leaking abstraction, we next consider null-terminated strings used in the C programming language, for instance. The following procedure can be used to concatenate two such strings:

```
char * strcat(char * c1, char * c2)
{
    int i, j;

    while(i = 0; 0 != c1[i]; i++);
    while(j = 0; 0 != c2[j]; j++, i++) c1[i] = c2[j];
    c1[i+1] = 0;
    return c1;
}
```

The logic of the operation is that first, we browse all the characters of string c1, and then copy all the characters of string c2 to its end. Moreover, it is assumed that c1 is large enough to host also the characters of c2, which is not explicitly expressed in the procedure but is an obvious built-in assumption.

From the functional viewpoint, using this kind of an operation appears perfect. However, from the practical viewpoint, the function is far from perfect, as in some cases the implementation of strings becomes visible to the user. A problem is that if we have to carry out one million concatenations to the same string, we

---

[1] The term 'leaking abstraction' has been used in this meaning at least by Joel Spolsky (2004). Also the example we use to demonstrate such abstractions originates from the same source.

will unnecessarily go through the same characters all over again as longer and longer strings adopt the roles of c1 and c2. While the execution would result in the correct outcome, the time needed for completing the execution would be considerably extended. By observing this from a completed, running system, one might be surprised since certain inputs would be slow to process, but by looking at the actual design, one would immediately learn the obvious problem of this implementation.

All non-trivial abstractions can be argued to leak to at least some extent (Spolsky 2004). For instance, let us consider the TCP/IP protocol that provides an abstraction of reliable communication; if the underlying communication infrastructure is terminally broken, there is no way the protocol can act in accordance to expectations. Similarly, although the SQL language is a powerful yet simple way to define database queries, some queries can (and usually should) be optimized by taking into account what takes place at the level of the implementation.

Fundamentally, programming languages which are commonly available in the mobile setting, such as C, C++, and Java, and their run-time infrastructures are also non-trivial abstractions. Therefore, they also have the potential to leak. In many cases, leaking abstractions of programming languages and infrastructures that are executed in mobile devices lead to problems in managing resources, memory consumption, and performance. Therefore, in order to compose programs where potentially leaking abstractions form a minimal problem, the programmer must have experience of working in a certain environment to create appropriate designs. Since expecting that all developers have experience is unrealistic, infrastructures have been introduced where the most obvious traps will be automatically treated, as well as tutorials and coding standards that aim at preventing the most obvious problems associated with leaking abstractions.

As already mentioned, mobile devices are restricted in terms of available resources. Therefore, in order to cope with leaking abstractions related to resources of the device, implied by the used programming languages and execution environments, the designer should understand what lies beneath the surface, because otherwise it is easy to use facilities of the language that are not well suited for such a restricted environment.

1.1.2 Allocation Responsibility

Programming systems in mobile devices can treat complexity in two fundamentally different ways. On the one hand, the responsibility can be given to the programmer, who then takes actions in order to manage resources, such as memory, disk space, or communication bandwidth. On the other hand, software infrastructure can be defined for handling the resources without revealing the details to the programmer, thus automating resource management from the programmer perspective. The above strategies can be considered as white-box and black-box resource management approaches in the sense that white-box resource management is visible to the developer in full, whereas black-box resource management hides its details from the programmer and aims at an automatic deallocation at an appropriate moment.

Programmer responsibility. Fundamentally, making programmers responsible for resource allocation results in a white-box approach to resource management that is relying on programmers who are able to carry out designs in a fashion where leaking of abstractions is not possible, or, at the very least, leaking is controlled by them. An obvious language where leaking of abstractions can be a problem in the mobile environment is C++, as in many cases features of the language require thorough knowledge of the underlying facilities and implementation techniques. Implementing programmer responsibility for potentially leaking abstractions can take many forms. On the one hand, one can define a coding standard that explains why certain types of designs should not be used, or are considered antipatterns, i.e., commonly applied solutions that bear some fundamental, well-known handicap (Brown et al. 1998). On the other hand, one can introduce a coding standard that defines design guidelines for managing cases where leaking of abstractions is considered most harmful or likely. Moreover, in addition to knowing the guidelines, the programmer should also understand what kinds of problems the guidelines solve and why. This usually calls for understanding of what happens at compilation and how run-time infrastructure works.

Infrastructure responsibility. A black-box approach to resource management means that the underlying programming infrastructure is supposed to liberate the developer from considering potentially leaking abstractions. Examples of environments that operate in this way include Java and C#, which both can be used for programming mobile devices. However, in practice the developer is still able to use the properties of the infrastructure better, provided that she knows how the infrastructure works and takes this into account when composing a design. For instance, being able to compose designs that do not overly complicate the work of a garbage collector in a virtual machine environment is helpful, as garbage collecting can seemingly stop the execution of a program for a short period of time in certain virtual machine environments. Thus, while the black-box approach hides resource management from the developer and the user, this abstraction leaks when garbage collection is performed, as its side-effects may become observable by the user.

To summarize, no matter whether the programmer or the infrastructure manages resources, the way programs are designed and written has an effect on their performance and resource consumption. In the following, we discuss the most common hardware-related issues that the developer is exposed to when designing applications for mobile devices.

1.2 Commonly Used Hardware and Software

To summarize the above discussion, programming languages and their run-time infrastructures can be considered as at least potentially leaking abstractions. Such leaks mean that the properties of hardware and lower-level software are revealed to an application programmer in some cases. Therefore, understanding their basics is a prerequisite for considering how applications should be designed for the mobile environment.

In the following, we give an overview to commonly used hardware and software facilities inside a mobile device, whose restrictions can be considered as the main technical contributors of mobility (Satyanarayanan 1997). The subsections address hardware, operating system concepts, application software, and the stack of software components that often forms the run-time environment of a mobile device.

1.2.1 Computing Hardware

The computing hardware of mobile devices can be expected to become more and more standardized. One important driver for this is the need to integrate all the subsystems into one chip. This saves development costs as well as energy consumption of the completed devices. Figure 1.1 illustrates the main elements that are significant within the scope of this presentation.

Processors and Accelerators

Fundamentally, a computer is a system that executes a program stored in memory. The processor loads instructions out of which the program is composed, and performs the tasks indicated by them. Instructions are low-level commands that discuss the execution in terms of hardware available in the system. For instance, loading the contents of a memory location to a processor's internal memory location, so-called register, adding the contents of two registers and storing the result in a third, and

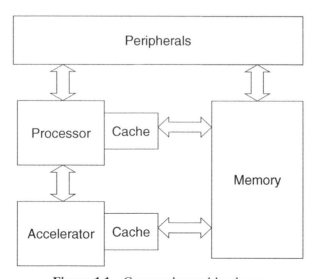

Figure 1.1   Commonly used hardware

```
int factor9() {
    int index, result = 1;
    for (index = 1; index < 10; index++)
    {
        result = result * index;
    }
    return result;
}
```

Figure 1.2   Sample source code

storing the value in a register to a particular location in memory, are commonly used instructions.

Processors have different instruction sets, i.e., primitive operations that processors are capable of executing, and their properties can vary. Probably the most commonly used main processor[2] in mobile devices follows ARM (Acorn RISC machine) design (Furber 2000). The ARM design is based on a 32-bit RISC[3] processor that is produced by several vendors, and which has been integrated into many more complex pieces of hardware, such as OMAP architecture by Texas Instruments. On the software side, for instance Symbian OS runs on top of ARM, although also other processor alternatives have been considered in the design of the operating system.

As an example, consider the following. The piece of code given in Figure 1.2 computes the factor of 9. When it is fed to a compiler, which in this case is a GCC compiler for Symbian OS, the resulting assembly output is as listed in Figure 1.3. Similarly to most, if not all, assemblers, the output includes individual references to locations in memory and to registers, as well as load and store instructions, which form the low-level representation of any program.

In addition to the main processor that is responsible for controlling the device as a whole, many mobile devices include different types of accelerators and auxiliary processors, which in terms of hardware can be considered similar to other processors, but they play a different role in the final system. They are used for coding and decoding of radio transmissions, and, more recently, to enable more sophisticated features such as three-dimensional graphics, for instance. Moreover, in some cases a proprietary phone implementation can be extended with an application processor that is to be used by additional applications without risking the functions of the proprietary phone. Further possible processors include an access processor (or a modem) that is dedicated for executing routines associated with telecommunications.

---

[2] In the terminology assumed here, the main processor is the processor that controls other parts of the system.

[3] RISC stands for Reduced Instruction Set Computer, where only a restricted number of relatively simple and quickly executable instructions are offered. In contrast, CISC, Complex Instruction Set Computer, refers to a more complex instruction set where also the execution times of different instructions can vary considerably.

```
@ Generated by gcc 2.9-psion-98r2 (Symbian build 540) for ARM/pe
.file "test.cpp"
.gcc2_compiled.:
.text
.align 0
.global factor9__Fv
factor9__Fv:
@ args = 0, pretend = 0, frame = 8
@ frame_needed = 1, current_function_anonymous_args = 0
mov ip, sp
stmfd sp!, {fp, ip, lr, pc}
sub fp, ip, #4
sub sp, sp, #8
mov r3, #1
str r3, [fp, #-20]
mov r3, #1
str r3, [fp, #-16]
.L2:
ldr r3, [fp, #-16]
cmp r3, #9
ble .L5
b .L3
.L5:
ldr r3, [fp, #-20]
ldr r2, [fp, #-16]
mul r3, r2, r3
str r3, [fp, #-20]
.L4:
ldr r3, [fp, #-16]
add r2, r3, #1
str r2, [fp, #-16]
b .L2
.L3:
ldr r3, [fp, #-20]
mov r0, r3
b .L1
.L1:
ldmea fp, {fp, sp, lr}
bx lr
```

Figure 1.3   Sample ARM code

Accelerators can be implemented using two different mechanisms, which are using multi-purpose hardware, such as a digital signal processor (DSP), and single-purpose hardware. Using the former is a more general design choice, as the hardware can be used for several tasks, whereas single-purpose hardware can only be used for the task it is designed for. However, a DSP is usually a bigger design block to be fitted in the device, and it requires more energy to run than a single-purpose piece of hardware. Still, in general, DSPs have superior power consumption to performance rate over general purpose processors in tasks that are well-suited for them. As an example, a study has shown that a typical signal processing task on a RISC machine (StrongARM, ARM9E) requires three times as many cycles as a C55x DSP while consuming more than twice the power (Chaoui et al. 2002). The downsides of DSPs are being solved by introducing more elaborated integration schemes for off-the-shelf hardware. Such integrated systems typically include a main processor, auxiliary DSP, memory, and additional interfaces. While systems exist where everything is integrated into a single chip, it is common that devices with new features include several chips, each of which is dedicated for a certain purpose.

A continuous trend is that an increasing number of features and improved integration techniques lead to more complex designs in the sense of architecture. In addition, many hardware and platform manufacturers have been emphasizing mechanisms related to power management, which also bear an effect on the complexity of the final design. Moreover, such features are becoming more important, as ready and use times of mobile devices are meaningful properties for the consumer.

When more and more advanced features that require more processing power are introduced, clock frequency of the main processor must be increased. Unfortunately, this leads to problems due to heat generation. Therefore, one can consider that the current approach, where there is a master processor and a small number of auxiliaries, may need to be reconsidered. As a solution, two contradicting approaches have been proposed. One is symmetric multiprocessing, which injects the complexity of programming to the operating system, and the other is the introduction of more and more specialized auxiliaries that manage their internal executions themselves. Presented in Figure 1.4, both approaches can be justified with solid arguments, as discussed in the following.

Symmetric multiprocessing (SMP). A system based on symmetric multiprocessing consists of a number of similar processors that usually work in intimate connection. A common implementation is to hide them in the same operating system core, which allows them to balance load between each other. This scheme gives an opportunity to save energy by shutting down some of the processors when the load is low, and keeping all processors active when processor-intensive tasks are being performed. Moreover, as all the processors can be hidden behind the same operating system interface and have similar characteristics, a programmer can be offered an abstraction that can be conveniently used. On the downside, multiprocessing can be a relatively complex solution, when assuming that fundamentally the device can actually be a mobile phone, which by nature is a very specialized system. Moreover, the adequateness of the processor-level granularity as the basis of energy management can be questioned.

Asymmetric multiprocessing. The other alternative, the introduction of multiple specialized pieces of hardware, also offers an increased performance. Furthermore, the design of individual pieces of equipment can be eased, as they all perform some particular tasks. However, allocating tasks to different pieces of equipment cannot be implemented in a straightforward fashion as with symmetric multiprocessing.

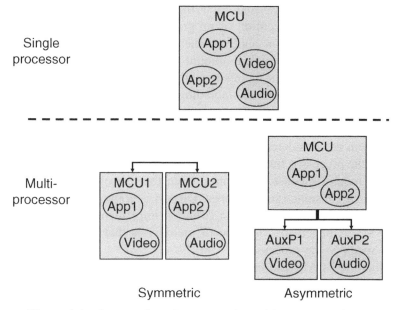

Figure 1.4   Symmetric and asymmetric multiprocessor schemes

Instead, the programmer may be forced to participate in the allocation for adequate results. As a consequence, different devices relying on different sets of auxiliary equipment must be supported. Moreover, some devices may implement some equipment with additional software running on the main processor. One possible result for this is to introduce abstractions that hide the complexity of hardware configuration in a standard fashion. Then, hardware support can be used whenever it is available, and software emulation will be used in devices that do not implement hardware acceleration. Still, a considerable development effort must be invested in the definition of interfaces that can be used with different implementations. Furthermore, standardization is required to enable interoperability. At present, the use of accelerators can be seen as a step in this direction.

Memory and Related Hardware

In current mobile devices, memory is usually internally constructed so that 32-bit memory words are used. Each word can be further decomposed to 4 bytes of 8 bits. In most cases, words are the main elements of memory in the sense that even if less than a word would be enough for a certain variable, a full word is still allocated in practice for performance reasons. There is, however, a possibility to allocate several variables to the same word, for instance. Unfortunately, this may result in degraded performance, as it is usually faster to access memory when respecting memory word boundaries. Respecting the word boundary is commonly referred to as (word) alignment.

Several types of memory are used. First, there is RAM (random access memory), which is used during the execution of programs for storing the loaded programs, and the state of the execution and variables related to it. This type of memory can be read and written. The typical amount of RAM in a mobile device has been increasing rapidly, with typical figures reaching up to 64 or even 128 Mb. Different types of RAM can be considered, including static RAM (SRAM) and dynamic RAM (DRAM). SRAM preserves its state but is unfortunately usually expensive, and it is commonly used only in memory that is to be accessed quickly, such as cache, which can be considered as an intermediate storage used for storing memory locations that the processor frequently accesses. DRAM, on the other hand, is based on transistors requiring constant attention from the rest of the system to preserve their state, which consumes some energy. DRAM is commonly used in the mobile environment, where probably the most common implementations rely on SDRAM (static DRAM). A benefit of this type of memory is that it can be run using the same clock speed as the processor.

Second, there is ROM (read only memory), which can be read but not rewritten. For instance, programs that are permanently stored in the device can be located in ROM. For execution, programs are usually first loaded to RAM for execution, although depending on the used chip set it is sometimes possible to execute programs directly from ROM using so-called in-place execution. In a mobile device, the amount of ROM can be 64 Mb or more, although flash memory, discussed in the following, can also be used for a similar role.

Third, many devices also contain permanent storage, although ROM and RAM would be sufficient for many purposes. The rationale is that if the battery is removed, it is helpful that the data stored in the device remains unaltered. Permanent storage can be implemented in terms of a hard drive or as flash memory which maintain their information even if power is switched off. Obviously, accessing disk is at least a magnitude slower than accessing RAM or ROM. Also physical characteristics of hard drives have couraged the use of flash memory, as it is not as prone to mechanical failures.

As already mentioned, flash memory is commonly used in mobile devices. Accessing flash memory can be implemented such that it is relatively fast to read from memory, but writing is usually slow. The reason is that it is possible to turn single bits from 1 to 0, but turning 0 to 1 can only be performed in groups of 64 kb for instance, depending on the hardware implementation. As a result, even a small change in a file can result in a complete rewrite of the whole file. Two types of flash memory are to be considered, NOR and NAND flash memories. NOR flash can be used as direct memory space, and as ROM or RAM. The latter is different from SDRAM in the sense that while an access to flash-based RAM can be slower, SDRAM requires constant attention of the processor to maintain the memory active, which consumes energy. In contrast to NOR flash, NAND flash behaves analogously to a hard disk, and requires loading of code to memory before running it. In comparison to other types of memory and hard disk, an additional restriction

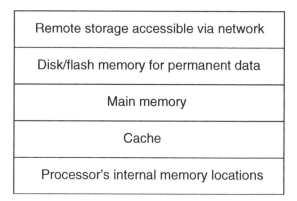

Figure 1.5 Memory hierarchy

is that only a very limited number of writes, say 100 000–1 000 000 rewrites, can be performed on flash due to its exhaustion. Similarly to hard disks, the use of flash requires the management of memory usage, which can consume a considerable amount of memory in a large flash file system (Chang and Kuo 2004).

The memory types discussed above create a hierarchy of memories (Figure 1.5), where the size of the memory and access times vary; at the bottom of the hierarchy (processor's internal registers or even cache) access is rapid and can be calculated in some nanoseconds, but only a limited amount of such storage space is offered. In contrast, accessing main memory takes place on the order of tens of nanoseconds, and accessing a value in disk is even slower, up to the order of tens of milliseconds. Furthermore, using some memory available in the network is slow, but a virtually unrestricted amount of memory can be offered.

In addition to the memory hierarchy, there is a special piece of hardware that is associated with memory, although it is not directly included in memory. The memory management unit (MMU, omitted from the figure for simplicity) is more conveniently implemented within the processor chip. The purpose of this piece of hardware is to enable using free locations in the memory for programs, disregarding their physical location. The MMU then manipulates the memory shown to the processor such that the memory image is simplified in the sense that programs usually appear to be in continuous locations in the memory even if they are distributed in the memory. Similarly, programs can be located in different parts of the memory during different execution instances of programs. MMU is also used for implementing memory protection enabled for processes, which we will discuss later in this chapter. Furthermore, MMU can also be used for implementing virtual memory, where some parts of the system are saved to disk in terms of memory pages when additional memory is needed, and when the pages are used again, they can be restored from the disk. As flash memory is not very convenient for such a purpose due to slow write operations and restricted number of write accesses, this scheme is not commonly implemented in mobile devices. However, the ability to freely

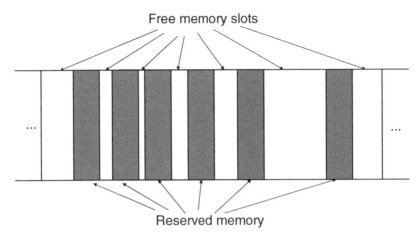

Figure 1.6 Fragmentation

locate programs in different parts of the memory as well as memory protection are appreciated, and therefore many mobile devices include an MMU. Still, there is no technical barrier to implementing a full virtual memory system.

Finally, two phenomena commonly associated with memory and its use are to be considered when composing programs for mobile devices. These are fragmentation and garbaging. The former means that memory is reserved such that some free memory remains unused between reserved blocks of memory (Figure 1.6). While fragmentation can in principle be handled by conjoining available free slices of memory to a bigger memory area, managing the lists of free and reserved slices can become a burden for the operating system, and superfluously consume precious memory and processing time. The latter refers to a situation where some data allocated to the memory can no longer be deallocated by the program that has allocated them, potentially reserving the memory locations until the execution of the program terminates, or, even worse, until the system is shut down.

Subsystems

In addition to the hardware described above, several types of auxiliary subsystems can be available, that require support from the hardware:

- Bluetooth is a short-range radio link that is used for cable replacement.
- Radio interface is used for communicating with the mobile network. Several bandwidths and protocols can be used, including for instance GSM, GPRS, WCDMA, and WLAN, to name a few.
- Keyboard (or touch screen) allows the user to input data and commands to the device.
- Screen enables the user to read the information stored inside the device.

- Additional memory for switching data between mobile devices using a memory stick that can be attached to a USB port or some other interface.
- Battery interface is used for managing the status of the energy source.

The above list is by no means exhaustive, and additional hardware is easy to imagine. In principle, anything that can be connected to a mobile device can form a new subsystem. Moreover, whether the connection is permanent or dynamic via a local area radio link, for instance, is becoming less important, as the capacity of connections is becoming comparable due to the increased coverage and bandwidth. This results in an option to use implementations where a number of devices cooperate to perform some higher-level tasks. At a lower level of abstraction, implementing this requires support from hardware in terms of I/O interfaces or the use of universal asynchronous receiver transmitters (UART), for instance.

In general, the characteristics of different subsystems are not directly visible to the application software running in the device as such but via different types of programming interfaces (or APIs, application programming interface), which makes it possible to run the same software even if the hardware is upgraded or modified in the design of a different type of device, which is common in the development of product families. However, lowest-level software must be adapted in order to benefit from the new hardware.

1.2.2 Low-Level Software Infrastructure

Kernel forms the core of an operating system. Fundamentally, it is a machine that is designed to manage the relationship of the underlying hardware and software that uses its resources.

For accessing hardware, kernels usually implement special modules called device drivers that can be used for creating a connection to the underlying subsystems and communicating with external equipment. The usual way to implement such drivers is that they send a request to the external subsystem, and the subsystem is responsible for serving the request. When the request is served, the subsystem responds with an interrupt that the kernel will acknowledge and serve with a corresponding interrupt handler, which contains the procedure for managing interrupts. Device drivers can be implemented in a layered fashion, where a physical layer handles the details of the actual hardware. A logical level can then be introduced for handling the parts that are common for all similar subsystems.

When considering the connection between the kernel and other software, the resources of the former are a necessity to create applications. Commonly used terms are processes and threads, which we will address in the following.

Processes can be considered as the units of resource reservation. This allows designs where the different resources allocated by programs are kept in isolation from one another. This, however, requires support from the hardware in the form of an MMU, which is usually included in processors used in mobile devices that can be extended with new applications. For devices that do not allow this and only include proprietary software, this may not be an option; instead, all processes can run in the same memory space, when they can interfere with each other's execution. Examples of resources that are managed by processes include memory in particular. In addition, threads that are run within the memory space of the process are resources owned by the process. In systems where threads could not be created separately, they were commonly associated with each other.

Threads can be taken as units of execution. Each thread can be considered as a conventional program that has a control flow, and some piece of code it executes. Threads belong to processes, and they can share resources and data structures with other threads inside the same process. When the thread that is being executed is changed to another by a special part of the kernel, the scheduler, the event is called a context switch. If the operating system can force a context switch even if the thread is not ready for it, the scheduler is called pre-emptive, and if the thread must explicitly pause its execution, the scheduler is referred to as non-pre-emptive. A pre-emptive scheduling policy is more flexible, but its performance ratio is usually recommended to be kept at maximum 70% as otherwise some operations that are important for the user can be delayed in favor of other operations. Moreover, context switching can also be executed repeatedly, if the performance ratio is increased. In contrast, for a non-pre-emptive scheduler, a fixed order of executions is usually defined, which is less flexible but whose performance ratio can be close to 100%. Two types of context switches can be considered; one takes place when a thread is changed to another but the process hosting the threads remains the same, and the other when also the hosting process is altered. The former is usually a lighter operation, but even this operation can be considered expensive in the sense of performance. The reason is that all the work performed for pausing one thread and selecting and starting the other is overhead.

1.2.3 Run-Time Infrastructure

In this subsection, we connect run-time infrastructure of programming languages used in the mobile setting to the facilities of the underlying hardware and operating system software.

Allocating Memory for Programs and Variables

Perhaps the most straightforward consumers of memory are constants, which can usually be saved in ROM. This saves some valuable RAM for other use.

For variables, two locations can be imagined. When a program is being run, its modifiable data must be stored in RAM, as the program can constantly alter the data. In contrast, when the execution is completed (or interrupted) data can be saved to disk for further use in some later execution. Saving data to disk is not simple, however. Disk (or flash memory) access can be slow, which can be problematic for the user. Furthermore, it can be difficult for the user to realize when all data has been saved and it is safe to turn off the device. Therefore, the design of programs

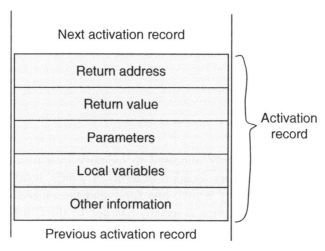

Figure 1.7 Activation record in a stack

should ensure timely saving of data in a fashion that is straightforward from the user perspective. In many mobile devices, this has been implemented using a practice where dialogs imitate transactions of database systems, offering for instance 'Done' selection for committing to the transaction in a fashion that is clear for the user.

Inside RAM, two fundamentally different locations can be used for storing variables, execution stack and heap. Execution stack is a memory structure that manages the control flow of a thread. Every method call made by a thread results in an activation record or a stack frame associated with the stack. The structure of an activation record is illustrated in Figure 1.7. This data structure includes information related to the management of the control flow, such as where to return once the execution of the called method is finished and what method made the call.[4] In addition, each activation record contains method parameters, the return value (if applicable), and variables that are local to a method, which in some systems are called automatic variables that are defined within the scope of the method. As the stack is intimately associated with the thread using it, the execution stack is usually local to a thread. In contrast to stack, which is structured in accordance to the execution of a program in terms of activation records, heap is an unstructured memory pool from which threads can allocate memory. Unlike memory allocated from stack, where the execution flow manages memory consumption, memory allocated from the heap is under the responsibility of the programmer. The accumulation of heap-allocated memory areas that are no longer accessible due to a programming error for instance is a common cause of garbaging; then, a programmer allocates memory, but never deallocates it.

---

[4] Exact details of activation records differ slightly in different systems. However, the principal idea remains the same.

The fashion in which the programmer composes programs determines where data structures are allocated. As an example on memory allocation from stack and from heap, consider the allocation of the following data structure:

```
struct Sample {
    int    i;
    char   c;
};
```

This data structure is most likely allocated to memory in a fashion where each variable takes a different memory location in terms of words, resulting in the consumption of two 32-bit memory words, one for i and the other for c. While the actual size of the memory allocation is usually handled by the compiler and associated run-time environment, its location in the program defines how data structures are partitioned between the stack and the heap. Automatic variables, i.e., those that are local to a method or procedure, are allocated from stack, and explicitly allocated data structures, using, say, malloc or new, are placed in the heap. Figure 1.8 represents two different ways to locate the above data structure to memory using stack and heap as the locations. As already discussed, the location that is allocated for a data structure affects the way memory is to be managed; stack-based variables are automatically created as the execution advances whereas heap-based variables require explicit allocation and deletion, assuming that no special infrastructure for garbage collection is introduced. Unfortunately, in some cases compilers perform

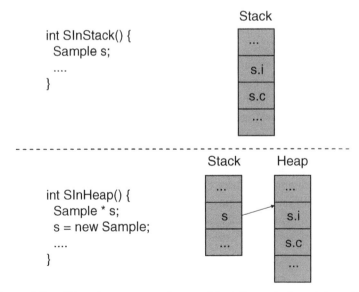

Figure 1.8 Allocating memory for an object from stack and from heap

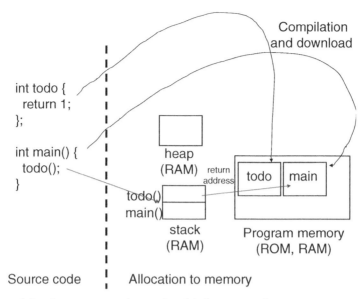

Figure 1.9   A program and associated infrastructural memory consumption

implicit allocations from heap even if a programmer assumes stack-based treatment. Such cases can lead to hard-to-trace errors that only become visible after an extended execution.

Programs are most commonly stored in ROM, flash, or disk, if such a facility is available in the device. When a program is run, it is usually loaded from its location in the storage to RAM for execution. This enables the use of updates for fixing bugs as well as the use of user-installed applications. Figure 1.9 demonstrates one way to allocate program and associated data in different memory locations.

For ROM-based programs, in-place execution can be used. This saves RAM, as there is no need to create an additional copy of the program into it. Usually this type of approach is used when a program is located in ROM, and it is beneficial to always run the program unaltered. For instance, in-place execution could be used for the introduction of a safety feature that enables (emergency) calls even if a mobile device is otherwise infected with a virus or some other malicious software. For flash memory, the type of flash used defines whether in-place execution can be used or not. As NOR flash can be used as direct memory space, programs stored to this type of memory are candidates for in-place execution, but because NAND flash behaves analogously to a hard disk, programs must always be loaded to RAM. There are also downsides to using in-place execution. First, it is impossible to upgrade the system using software modules that are downloaded after the installation. Second, it is not possible to use compression or encryption techniques, as the code must be executable as such. Therefore, in-place execution is becoming less favorable than it was in mobile devices where the option to download new applications was not available.

Generated Run-Time Elements

As already discussed, all programs and variables in them are allocated to memory locations when a program is executed. In addition to the basic cases, more sophisticated software structures introduce additional memory consumption. Such structures include infrastructure for inheritance and virtual machines, which we will discuss in the following.

In association with inheritance, an auxiliary data structure called a virtual function table is commonly generated. Its goal is to manage dynamic binding (Figure 1.10). When referring to a method during the execution of a program, dynamic binding does not refer to the method directly, but via a reference to the methods of the associated class and an offset that identifies the exact method to be called. Effectively, this implies that for all classes where dynamic binding can be used, a table is created where the cells of the table refer to different functions of the class. Then, when a program runs to a virtual function call, object type is used to select the right virtual function, and an offset included in the generated code is used to select the right function. As the outcome, all classes that potentially use dynamic binding require additional memory for the virtual function table, and, furthermore, add one extra memory word in each object for the type tag, which in reality often is a reference (or pointer) to the right virtual function table.

A virtual machine can be taken as a processor (or interpreter) implemented with software. This eases porting of programs implemented on top of the virtual machine, as the same (or similarly instructed) virtual machine can be used in different hardware environments. Fundamentally, two types of virtual machines can be considered. One is such that an interpreter is used that executes programs by simply interpreting each instruction as such. The other approach is that once the program to be run is fed to the virtual machine, the virtual machine compiles it into

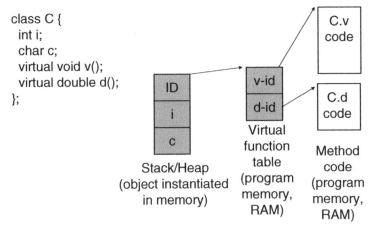

Figure 1.10   Virtual function table

an executable form that is closer to the actual machine code and can therefore be executed faster. Also a combination of the two is possible. Then, some commonly executed parts of the program for instance are compiled, whereas parts that are only executed seldomly are interpreted every time. This technique is referred to as hotspot compilation. A further issue worth considering is that since programs can be interpreted as data that are fed to a virtual machine, the machine can impose additional restrictions on programs. Moreover, it can control resource usage of programs, and manage loading of programs, an important facility to centralize in the mobile setting, to execution. A sample virtual machine, which reflects the features of a Java virtual machine, is illustrated in Figure 1.11 (Hartikainen, 2005). The roles of the different components of a virtual machine are listed in the following.

- Class loader is a component responsible for loading programs, given in terms of classes in our reference system.
- Run-time data is a container for loaded programs. It includes elements similar to those used in native executions for the purposes of the virtual machine.
- Execution engine contains a scheduler, which is responsible for selecting a thread for execution, a memory manager, which ensures that no illegal memory references are made and that memory is deallocated when it is no longer needed, and a bytecode interpreter, which executes the actual programs.

Garbage collection, i.e., freeing of resources that are still reserved but are known to be abandoned, can be implemented in two different ways using a virtual machine. Usually, a simple way is to implement garbage collection in cooperative (or stop-the-world) fashion, which stops the application while garbage is being collected. Therefore, only the garbage collector has access to data, and it can perform its task without a risk of modifying the same variables as the application at the same time. In contrast, parallel garbage collection can also be implemented, but this requires

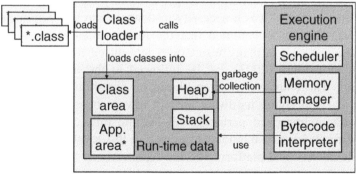

* App. area includes e.g. Program Counter

Figure 1.11   Elements of a Java virtual machine

a more complex design because the garbage collector and the application being executed operate on the same data structures. In general, garbage collection is a wide topic with dedicated books (Jones 1999), and its detailed introduction at the level of individual algorithms, such as reference counting, mark-and-sweep, heap compaction, or more recent generational garbage collectors, falls beyond the scope of this book.

1.2.4 Software Stack

Building on top of kernel and other low level facilities, software used in mobile devices can be characterized using three different categories of software: applications and related facilities, middleware, and low-level software. The reason for separating these layers is that it is often desirable to consider that such levels of abstraction can evolve independently of each other, i.e., it is preferable that applications are independent of middleware version, and that middleware in general is not tied to a particular version of low-level software but can be used over several generations. Furthermore, maintained and relatively static interfaces are often defined to separate the layers. In the following, we address these categories one by one.

1. *Application-level software* is about the development of meaningful applications for an end user. In mobile devices, common application development rules of workstation software apply to a great extent, as it is often the intention of the device manufacturer to allow the introduction of additional applications. Implementation of other software components can benefit from using the provided interfaces of the lower-level components, but their implementations should not be relied on because different versions of software can be used in different devices. Moreover, a characteristic property of application-level development is usability, as otherwise users may reject the application.
2. *Middleware software* offers facilities that ease the development of applications. This often includes libraries such as support for using certain communication protocols. While an application developer is usually only using already existing systems, in some cases it is a necessity to define some new application-specific components. Then, it is common that the implementations must follow some predefined interfaces. In many ways, rules introduced for communications programming by Sridhar (2003) can be followed in the implementation of such components. The emphasis is often placed on portability in the sense that the details of the underlying hardware are not always benefited from even if this would result in improved performance. Despite this, characteristic properties include performance and memory awareness at this level of abstraction in practice despite device- and platform-specific features.
3. *Low-level software* covers kernel and device drivers, and virtual machines when applicable. Usually all such software is fixed by the device manufacturer. When developing such software, guidelines of Barr (1999) can be benefited from due

to the fact that in many ways the development of low-level software resembles the development of deeply embedded systems, with close connection to the underlying hardware. However, even at this level of abstraction, some precautions can be taken for reusability in future hardware environment, for instance. A characteristic property at this level is hardware awareness in general, but in a fashion that does not impose too harsh restrictions. In general, software associated with this level of abstraction is managed by a device manufacturer or platform developer.

To summarize the above discussion, the lower one goes in the sense of abstraction, the more one must understand the properties and restrictions of the platform and the hardware environment. Unfortunately, even at the application level, some properties of the environment remain visible, making applications a leaking abstraction. Moreover, when aiming at applications that can be used in multiple devices, also device variance can be considered as a main issue in the development.

A further problem results from the fact that the levels are seldom static but evolve as more and more devices are implemented on top of the same set of software components, often referred to as a platform.[5] In this setting, it is common that upgrades take place and introduce new features to the platform as more hardware and processing facilities are introduced, as well as standardization advances. If performed in a careless fashion, the introduction of a new version of some low-level feature can lead to invalidation of some applications. Therefore, extra care should be paid for maintaining (binary) compatibility when composing new versions of middleware and low-level software. Otherwise, the platform can corrupt to a collection of devices in which one cannot run the same software, but must make device-specific modifications for all software that takes into account specifics of devices that are to be used for running the application.

For practical reasons, the development of a platform and the development of an application may have different concerns. For instance, for a particular application, many concerns related to variability to different types of devices – essential aspects of platform development – become irrelevant, if the application is targeted to only one type of device. Similarly, tailoring applications to run on as many platforms as possible often becomes relevant for an application developer only when the first running version is available, whereas tailoring a platform such that it will run all the applications constructed in accordance to some guidelines must be taken into account at the very beginning of the development as a compatibility requirement.

1.3 Development Process

The development process of a piece of software intended for a mobile device differs from that of conventional workstation software. The main reason is that first,

---

[5] Strictly speaking one can consider that all the above levels constitute different platforms: OS platform, middleware platform, and application platform. Also other categories have been proposed.

the developed software is often tested in the development environment, usually a PC, with an emulator. Only when the software runs is it downloaded to an actual mobile device for testing, as depicted in Figure 1.12. The development workstation is commonly referred to as a host, and the final execution environment as the target.

The software tested with the emulator and the software downloaded to a mobile device can be identical, but they need not be. If the same programming infrastructure is available in both environments, like a common virtual machine for instance, the software can obviously be the same, as the case can be with Java. However, if there is no infrastructure that would enable the use of the same software, different compilations are needed. The process where a workstation is used to compile an executable which is downloaded to a different system is referred to as crosscompilation. While it is possible to emulate the low-level behavior of a system, differences in available environment sometimes lead to early use of the actual device even in the development phase.

Obviously, being able to run the same version in the development and the final execution environment eases debugging, as only one tool-chain is needed. In the best case, downloading the application to the final environment is trivial. However, if cross-compilation is needed, the phase where software runs in the emulator can be only the beginning of debugging and testing activities that are necessary until the program is completed. For instance, there can be additional requirements regarding compilation, if the mobile device does not support all the features of the emulator. For example, older versions of Symbian OS (before v.9.0) were unable to handle

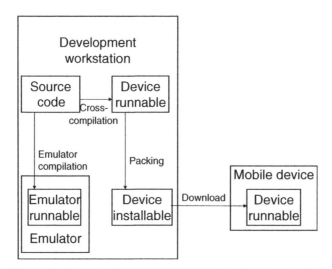

Figure 1.12   Development process and software download

global variables in dynamically loaded libraries in devices,[6] but forced one to use a platform-specific feature called thread local storage (TLS), where thread implementation was piggy-backed with global variables (Tasker et al. 2000). However, this problem only arises in the cross-compilation phase, as it is related to the final execution environment, not to C++ that is used as the programming language or emulator, which runs on the development workstation and can deal with global variables without problems.

In addition to the actual cross-compilation, many environments require the use of installation files. In addition to plain compiled programs, such files can often be extended with additional data, which can be auxiliary data files, or sound or graphics extensions for a certain program, for instance. When a package containing all these files is fed to the installer application residing on the mobile device, the installer then unpacks the files and places them to convenient locations as defined by the package.

1.4 Chapter Overview

The chapters of this book introduce the main concerns of the design of software for mobile devices. Unlike many other introductions, we have organized the presentation in accordance to concerns, which are memory usage, concept of applications, modularity and available mechanisms, concurrency, generic resource management, networking, and security. Individual chapters and their contents are listed in the following.

Chapter 1 has introduced the basics of mobile devices in terms of underlying hardware and software. The goal of the chapter is to introduce the basic concepts on top of which further chapters will build on.

Chapter 2 discusses memory management, which a designer must master in the mobile setting. As memory is one of the restricted elements of a mobile system, it is important to understand the basic principles of managing its use. The chapter introduces some design patterns, i.e., reusable design decisions that solve certain problems in a predefined context (Gamma et al. 1995), for memory-aware software, and gives examples on memory management in mobile devices. Similar topics have already been addressed by Noble and Weir (2001), although their focus is been wider.

Chapter 3 introduces the concept of an application. While in principle, all software can be implemented from ground up, mobile software platforms usually introduce a prescribed architecture for applications. Furthermore, another important issue is packaging applications for delivery to a device.

---

[6] The reason for not using global variables in dynamically linked libraries lies in the fact that at least one additional memory page, i.e., memory allocation unit, would be reserved for the library if global variables were allowed. This would lead to a considerable overhead.

In Chapter 4 we discuss the use of dynamically linked libraries, and demonstrate how available implementation techniques become visible to a programmer in a mobile environment. Again, the purpose is to address how the selected design and implementation decisions are unveiled to programmers.

Chapter 5 introduces the most important mechanisms of concurrency applicable in the mobile setting. The chapter addresses issues like whether to use threads, which are a common technique when programming desktop systems, or to use other means of serialization, which can offer less resource consuming solutions. Moreover, the chapter considers reusability of such designs.

Chapter 6 integrates concurrency with the management of different resources that must be handled in a mobile device. The chapter also discusses problems associated with the fact that it is not uncommon that even seemingly similar devices can include some differences in their hardware, which can sometimes lead to complications in the development of associated software.

Chapter 7 defines how a mobile device can interact with networks. The goal is to describe how mobile devices can be used in a networking application, and how software residing in a mobile device should be designed, and general topics associated with distributed systems and their implementation are mainly overlooked. Furthermore, the chapter also discusses two cases, one on using Web Services with a mobile device, and another on ad-hoc networking over short-range wireless protocols using Bluetooth as an example.

Chapter 8 is dedicated to security properties of a mobile platform. The chapter discusses both design patterns for secure designs as well as approaches adapted in existing platforms. The rationale of locating this important topic towards the end of the discussion is that security is an issue that contributes to all the previous techniques in a fundamental fashion. Without first discussing the basic implementation, it would be impossible to introduce a suitable security concept for the different issues.

In each chapter, we will use mobile Java (Riggs et al. 2001; Topley 2002) and Symbian Operating System (Edwards et al. 2004; Harrison 2003; Tasker et al. 2000) as examples on real-life mobile platforms. The reason for using these systems is that in many ways they contradict each other. In mobile Java, the underlying mindset is that it can be introduced as an external facility to all proprietary phones, and that the run-time infrastructure will adopt a major portion of the complexity of dealing with scarce resources. In contrast, in Symbian OS, the complexity is explicitly made visible to the programmer, together with certain patterns and idioms that are to be used when dealing with resources. When used with care, this in principle yields improved performance, because the whole power of the device's hardware can be used.

In addition, we also give some exercises on the topic discussed in the chapter. However, the purpose is not to focus on these particular platforms but to maintain a more abstract approach, but examples merely characterize some design solutions made in real platforms.

1.5 Summary

- Programming of mobile devices is fundamentally the same as programming of desktops. However, design flaws are more prone to cause problems, as resources are scarce in mobile devices. Moreover, available facilities are usually not as advanced as when programming a workstation, but developers must sometimes rely on practices more commonly associated with programming of embedded systems.
- Understanding the characteristics of the underlying hardware and software infrastructure is a practical necessity for dealing with leaking abstractions in the mobile setting.
- While there are several types of devices, they share the same basic architecture.
  - Processor, a hierarchy of memory facilities, and auxiliary subsystems for interacting with the environment and enhanced features.
  - Operating system, middleware, and applications, whose reliability requirements may differ. This has implications for software development as well.
  - Ability to reuse existing infrastructure as a common platform in multiple devices is preferable.
- A fundamental design decision of future mobile devices is whether to use symmetric multiprocessing versus a collection of specialized pieces of hardware.
  - Symmetric architecture eases the development of applications in the sense that all facilities appear similar to the programmer.
  - Asymmetric architecture can be more specialized in its resource use, which often hardens design and adequate use of available facilities.
- Run-time software infrastructure can introduce overhead; examples of such overhead include virtual function table used for implementing dynamic binding associated with inheritance, and virtual machines.

1.6 Exercises

1. What solutions would be applicable for defining an abstraction that does not leak for strings? Which programming infrastructures (languages, operating systems, etc.) use them?
2. What properties of applications have potential for leaking in the mobile environment, assuming that a programming language familiar to you is available? How should the developer address these properties?
3. What differences can you find in your own software programs when considering the way in which they use stack and heap? Why?
4. Consider a calendar application running in a mobile device. Which services does it use from low-level, middleware-level, and application-level components?
5. Which features of programming languages, such as C++ or Java, can be problematic when programming mobile devices? What kinds of coding rules or idioms

could be given for a developer to ease the development, assuming that for instance memory usage and resource management are to be controlled better?

6. In-place execution enables the execution of programs directly from ROM. What rationale is there not to implement all features of a mobile device using this technique?

7. What functions could be separated from the main processor to be executed by some auxiliary unit? What would be a reasonable level of abstraction to offer to the application developer?

8. What parts of the hardware should be standardized for a mainstream workstation operating system to be assumed as the platform for mobile applications? What would this mean for application developers?

2

Memory Management

2.1 Overview

Memory is an important resource for any computing system. However, memory management is different from managing some other resources in the sense that there can be no single module that would be responsible for memory management in isolation of the rest of the system. Instead, memory use is necessarily tangling everywhere in a program. Furthermore, when considering programming mobile devices memory is a critical resource, because in an attempt to keep the cost of the device low, manufacturers include only a restricted amount of it in devices although all the running programs are competing for it. Moreover, in addition to forming a considerable cost factor, memory chips also consume some power, the amount of which depends on the amount of memory included in the device. Therefore, many devices are limited with respect to their memory also due to this reason.

At the same time, memory is a crucial resource whose use cannot be abstracted away. Instead, programmers essentially define how memory is used in programs, as a common goal of language design has been to enable explicit allocation of memory in many imperative languages, such as C and C++. Furthermore, the way programmers write their programs has a large contribution to the memory usage of the program. In fact, in some mobile platforms, almost all programs show signs of preparation for the case where the program runs out of memory or some other resources, making memory consumption a cross-cutting problem when composing programs.

2.2 Strategies for Allocating Variables to Memory

As already discussed, the programmer defines how variables are allocated to different memory locations, enabling the use of different memory areas. While the stack and the heap are in principle just memory areas, the fact that the responsibility for their management lies with the run-time infrastructure and the programmer, respectively, leads to practical differences. Additional considerations should be paid to the fact that when using the heap, sharing of data is easy and natural, whereas with stack-based variables one should use references with care, if at all, because as the execution proceeds the stack increases and decreases, and may overwrite referred data. Furthermore, whether a variable is statically allocated to a permanent location or dynamically allocated from stack or heap is commonly visible at the level of programming languages used when composing programs to mobile devices, which in fact implies that the designers of the language have wanted that the programmer designs how a program uses memory. This forces the programmer to design the allocation.

In the following, we discuss the basic strategies for using global variables as well as selecting the stack or the heap for allocation of a variable.

2.2.1 Static Allocation

As already hinted above, perhaps the simplest case of allocation is static allocation. Then, a variable is statically allocated to a certain location in memory, and all references to it always address this location. What makes the situation simple is the fact that the variable remains allocated throughout the execution of the program, and the memory locations used by it cannot be deallocated.

If a programmer wishes to allocate a variable from memory in a static fashion in a method or in a class, this can be explicitly requested. For instance, the following example introduces such a variable (Koenig and Moo, 2000):

```
int * pointer_to_static()
{
    static int x;
    return &x;
}
```

While the way the variable is allocated suggests that it is an automatic variable that is local to the function, it is in fact allocated in the memory in a similar fashion to global variables, which are also statically allocated variables. The reason for introducing the variable in the above fashion rather than as a global variable is the scoping; now only restricted visibility to the variable is provided, as it does not pollute the program's global variable space.

In the object-oriented setting, the Singleton pattern introduced by Gamma et al. (1995) is sometimes used as a mechanism for static allocation. There, the goal is to define a single instance (or a few instances) of a class, and make it known to all subsystems. In practice, this is accomplished by allowing the class to be responsible for the creation of the object.

2.2.2 Stack

As a rule of thumb, transient objects, i.e., those that live only a limited period of time, are to be stored in the stack. The idea is that if an object lives a short period of time in any case, it is easier to allocate it in stack, because the object will be deallocated automatically as the execution advances. Furthermore, using stack memory is usually already reserved, and searching for a suitable memory area need not be performed upon allocation. This issue further advocates the use of stack for transient objects.

On the downside, automatic allocation and deallocation of variables creates a potential problem for using references to variables in the stack, as the variables can be erased and replaced by some other variables. Assuming that references are always made from later activations, this can never cause a problem. In practice, however, ensuring this in design and in particular in the maintenance phase can turn out to be hard.

As an example, assume that a programming error occurs, and the following function is written instead of the above example of using static variables:

```
int * pointer_to_int() // WARNING: This is a negative example.
{
    int x;
    return &x;
}
```

Now, the returned reference points to a location in the memory, that is potentially overwritten by any later functions that will be called by the program (Figure 2.1). Obviously, the resulting execution easily becomes unstable, as the value of the memory location can be altered seemingly nondeterministically as the program advances. Furthermore, also methods that assign to this variable are dangerous,

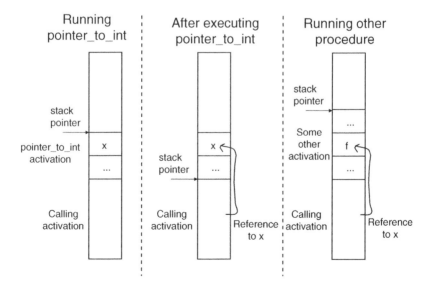

Figure 2.1 Erroneous use of stack-based variable

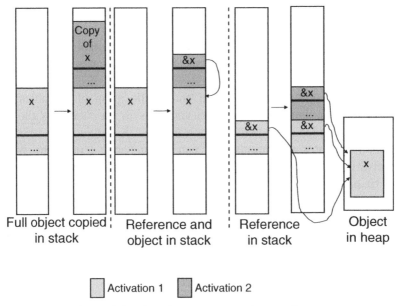

Full object copied ┊ Reference and ┊ Reference   Object
      in stack   ┊ object in stack ┊ in stack   in heap

☐ Activation 1   ◼ Activation 2

Figure 2.2   Copying versus passing a reference

since they can modify any data that is located in the stack, including parameters, local parameters, and return addresses. This often results in hard-to-repeat errors and complex debugging tasks.

An additional limitation of stack is that it is possible to compose a program where all data is accidentally copied when making a method or function call, which may result in exhausting the stack. Therefore, it is usually better to allocate big objects from the heap, and only pass a reference as the parameter, which results in less memory consumption (Figure 2.2). Both the stack and the heap can be used as the host for the referred data structure. Disregarding the host, this creates a situation where there are several ways to alter the data in the object, a situation referred to as aliasing. In order to ensure that the object is not accidentally modified via a reference, const definition can be used where applicable. Furthermore, following Copy-on-Write practice, where any write access automatically creates a new copy, can often be used to maintain the situation simpler.

2.2.3 Heap

All data structures whose size or structure can be altered dynamically must be allocated to the heap. The rationale is that since their size cannot be known in advance, it is impossible to reserve enough space from the stack.

Another reason that sometimes leads to using the heap rather than the stack is that the object must live despite the phase of the program. In other words, if the role of an object is global, then memory for it should be allocated from heap.

As already discussed, inside methods such variables can be declared static, which essentially make them global. Moreover, if large objects are to be allocated, they can be allocated to the heap to avoid exhausting the stack, whose size can be limited for a program.

If there is no automation regarding garbage collection in the run-time system, it is a responsibility of the caller of this function to ensure that memory is released when the data structure is no longer needed. At the same time, it is a good practice to reset the reference to the area, so that it will not be accidentally used.

It is also important to remember that using the heap can be slower than using the stack, as the running program may have to search for a suitable memory area. Common run-time implementations are based on a list of references to available memory areas hosted by the run-time environment, from which the suitable area is selected. Another possible implementation is to use a stack-like pool of memory, but when objects are allocated and deallocated, it is common that the pool becomes fragmented and a list to free memory areas is still needed. Moreover, in the worst case, none of the areas can be selected, and they are either to be merged to create a larger memory area or some more memory is queried from the operating system, which in turn is an even slower operation, involving context switching and all the associated complexity.

2.3 Design Patterns for Limited Memory

When composing designs for devices with a limited amount of memory, the most important principle is not to waste memory, as pointed out by Noble and Weir (2001). This means that the design should be based on the most adequate data structure, which offers the right operations. For instance, one should not use a two-directionally linked list if one direction would be enough. In addition to this basic principle, a number of other considerations are related to memory management, where the objective is to use memory such that its implementation leads to minimal leaking of abstraction, or, where applicable, the underlying implementation provides improved properties, like for instance better performance if support can be gained from cache.

In the following, we introduce some design patterns created for help in designing small memory software. Many of the patterns discussed in the following have been introduced in more detail by Noble and Weir (2001), but in a more generic setting. Here, we focus on their application in the design of programs running in mobile devices.

2.3.1 Linear Data Structures

In contrast to data structures where a separate memory area is reserved for each item, linear data structures are those where different elements are located next to each other in the memory. Examples of non-linear data structures include common implementations of lists and tree-like data structures, whereas linear data structures can be lists and tables, for instance. The difference in the allocation in the memory also plays a part in the quality properties of data structures.

The principal rule is to *favor linear data structures*. Linear data structures are generally better for memory management than non-linear ones for several reasons, as listed in the following:

- *Less fragmentation*. Linear data structures occupy memory place from one location, whereas non-linear ones can be located in different places. Obviously, the former results in less possibility for fragmentation.
- *Less searching overhead*. Reserving a linear block of memory for several items only takes one search for a suitable memory element in the run-time environment, whereas non-linear structures require one request for memory per allocated element. Combined with a design where one object allocates a number of child objects, this may also lead to a serious performance problem.
- *Design-time management*. Linear blocks are easier to manage at design time, as fewer reservations are made. This usually leads to cleaner designs.
- *Monitoring*. Addressing can be performed in a monitored fashion, because it is possible to check that the used index refers to a legal object.
- *Cache improvement*. When using linear data structures, it is more likely that the next data element is already in cache, as cache works internally with blocks of memory. A related issue is that most caches expect that data structures are used in increasing order of used memory locations. Therefore, it is beneficial to reflect this in designs where applicable.
- *Index uses less memory*. An absolute reference to an object usually consumes 32 bits, whereas by allocating objects to a vector of 256 objects, assuming that this is the upper limit of objects, an index of only 8 bits can be used. Furthermore, it is possible to check that there will be no invalid indexing.

Linear memory allocation often requires that memory is reserved in advance, at least partly. It is then possible to use the already reserved objects later in the user program. This gives rise to some related basic principles to consider.

2.3.2 Basic Design Decisions

In the following, we introduce some basic principles helping in using linear data structures. The purpose is not to introduce a complete checklist, but rather offer some examples on how linear data structures can be benefited from when composing designs.

Allocate all memory at the beginning of a program. This ensures that the application always has all the memory it needs, and memory allocation can only fail at the beginning of the program. Reserving all the resources is particularly attractive when the most important or mandatory features like emergency calls, for instance, are considered, for which resources must always be available. In general, this type of an approach is best suited for devices that have been optimized for one purpose, and it cannot be generally applied in smartphones except only in some restricted special cases.

Allocate memory for several items, even if you only need one. Then, one can build a policy where a number of objects is reserved with one allocation request. These objects can then be used later when needed. This reduces the number of allocation requests, which leads to a less complex structure in the memory. The approach also improves performance, as there will be fewer memory allocations, and cache use is improved.

Use standard allocation sizes. With a standard allocation size, it is easy to reuse a deallocated area in the memory when the next reservation is made. As a result, fragmentation of memory can be prevented, at least to some extent.

Reuse objects. Reusing old objects might require using a pool of free objects. This requires some data structure for managing free and used data structures. This implies that the programmer actively participates in the process of selecting object construction and destruction policy in the design.

Release early, allocate late. By always deallocating as soon as possible the programmer can give more options for memory management, because new objects can be allocated to the area that has just been released as well. In contrast, by allocating memory as late as possible, the developer can ensure that all possible deallocations have been performed before the allocation. In particular, one should ensure that objects occupying a large amount of memory are deallocated before allocating new objects. The reason is that in many implementations, heap gives the first suitable memory area, or, in a stack-like implementation, on one end. Then, when large objects are deallocated before allocating others, fragmentation can potentially be prevented, or at very least its effect can be lessened.

Use permanent storage or ROM when applicable. In many situations, it is not even desirable to keep all the data structures in the program memory due to physical restrictions. For instance, in a case when the battery is removed from the device, all unsaved data will be lost. For such situations, it is advisable to introduce the custom to save all data to permanent storage as soon as possible. This can be eased with a user interface that forces the user to commit to completing an entry to calendar or contacts, for instance. A similar fashion can be derived for static data, such as dynamic library and application identifiers or strings used in applications. Furthermore, even if there is no risk of losing data, it may be beneficial from the memory consumption point of view to write large, seldom used objects to permanent storage, so that the device's memory is preserved for more important data.

Avoid recursion. Invoking methods obviously causes stack frames to be generated. While the size of an individual stack frame can be small – for instance, in Kilo Virtual Machine (KVM), which is a mobile Java virtual machine commonly used in early Java enabled mobile phones, the size of a single stack frame is at least 28 bytes (7×4 bytes) – functions calling themselves recursively can end up using a lot of stack, if the depth of the recursion is not considered beforehand. However, this seldom is a problem in applications that are meaningful in mobile devices in practice, as they rarely perform computing that would require deep recursion.

2.3.3 Data Packing

Data packing is probably the most obvious way to reduce memory consumption. There are several sides to data packing, however. In the following, we discuss some alternatives that may bear significance. However, it should be emphasized that selecting the right data structure as the basis of an implementation is usually vastly more important than packing in the majority of practical designs.

Consider word alignment. Due to word alignment, it is possible that data structures are not optimally located in the memory. For instance, it is impossible to allocate variables to memory such that a 32-bit variable would be allocated in between two 32-bit words. Therefore, the data structure

```
struct S {
    char c1; // Actually a boolean.
    int i;
    char c2;
};
```

cannot be allocated to two memory words, but three words are required: c1 reserves 8 bits from the first word, then i must be allocated to the next full word in memory, and finally c2 must be allocated to the third full word. The rationale is that if the compiler automatically optimized the layout of data structures in the memory, complications might result if the data structure was used as a mapping to a specific piece of hardware to the program's memory space that only makes sense in this particular order. This is a common case in low-level programming where the abstraction of data types sometimes reveals its implementation. By changing the data structure to

```
struct S {
    char c1; // Actually a boolean.
    char c2;
    int i;
};
```

the developer can manage with two memory words, as c1 and c2 fit in the same word. However, optimization for performance may prevent this, as it can be slower to address variables that are located in the middle of a memory word. Moreover, a way to instruct the associated compiler is also needed. Further improvement can be gained by using only one bit for the boolean value, assuming that there would be some other data to include in the saved bits. The situation is illustrated in Figure 2.3, but with only 16-bit words due to practical reasons.

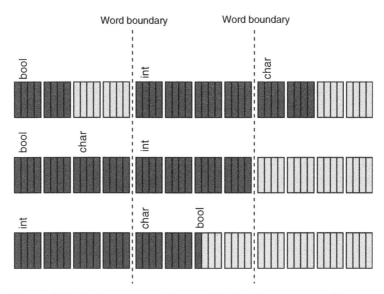

Figure 2.3 Saving memory by considering data structure alignment

Use compression with care. In addition to considering the data layout in memory, there are several compression techniques for decreasing the size of a file. Since packing as a technique is simple it can be applied to both code and data structures, at least in principle. In practice, however, one should be careful with packing, because opening packed files can impair performance. The same applies to the generation of compressed data, if a compressed format is used as the internal representation by the device. Noble and Weir (2001) introduce three different compression techniques:

- *Table compression*, also referred to as nibble coding or Huffman coding, is about encoding each element of data in a variable number of bits so that the more common elements require fewer bits.
- *Difference coding* is based on representing sequences of data according to the differences between them. This typically results in improved memory reduction than table compression, but also sometimes leads to more complexity, as not only absolute values but also differences are to be managed.
- *Adaptive compression* is based on algorithms that analyze the data to be compressed and then adapt their behavior accordingly. Again, further complexity is introduced, as it is the compression algorithm that is evolving, not only data.

In practice, it is not uncommon to use table compression and difference coding as a part of or with adaptive compression.

Use efficient resource storage format. Based on Hartikainen (2005), one must take special care to ensure that images, sounds, and movies are stored in the most efficient way as it is very likely that they consume a considerable amount of memory. The same applies to format, compression, bit depths, sampling rates, and resolutions. It is possible to save some space by combining many image files into one image file. This way overhead from file headers can be minimized. If images are similar in content then also compression might achieve better results. Finally, the most savings from combining images can be achieved when there are lots of small images.

2.3.4 Discussion

The above patterns for memory management are not problem-free. As pointed out by Noble and Weir (2001), when improving a certain property of software design, it is common that some other part is downgraded or compromised. In general, assuming that the appropriateness of memory-consciousness is acknowledged, taking information on implementation techniques into account can lead to smaller memory consumption. However, a downside is that in many cases other properties of the design can be harmed. This can involve at least the following aspects.

- *Increased minimal memory usage.* In anticipation of larger data amounts, it is possible that the minimal amount of memory needed for data is more than in a simplistic implementation.
- *Decreased flexibility.* When assumptions about the underlying implementations are made, it is possible that some hardware configurations are invalidated.
- *Downgraded performance.* Using some form of compression can lead to decreased memory use. However, encoding and decoding of data is harmful from the performance perspective. In the simplest form, even packing a lot of information into the same memory word can lead to downgraded performance.
- *Longer initialization and shutdown sequences.* As in some cases it is possible to handle some of the operations associated with memory at the beginning or at the end of the program rather than in the middle of the execution, startup and termination of the program can become slower. Moreover, when some other program with more modest requirements needs to execute the same initialization sequence, the resulting execution can be considered unnecessarily slow.
- *Potential unintuitiveness in designs.* When an experienced designer composes a design in a memory-aware fashion, some of the decisions can be unintuitive to less experienced developers. Over a period of maintenance, this can lead to a system that has all the downsides of different designs the developers have incorporated in the system, but few of the benefits. This kind of a decayed system is increasingly difficult to maintain, and its properties are often hard to recognize.
- *Impaired reusability.* The more one addresses particularities of a certain design problem in the solution, the less likely it is that the same solution could be used again in another context.

To summarize, it is a necessity to balance between the different requirements addressing the properties of designs. In practice the ability to compose designs where bargaining with different resources is needed, as it is common that the requirements change over time. For instance, when composing the first version of a certain system, it is seldom subjected to tight timing and sizing budgets, but flexibility is an important requirement. However, over time, it often becomes obvious that by reworking some parts of the design some memory can be saved and performance can be improved at the cost of flexibility that was considered desirable in the beginning, but has turned out to be superfluous later in the development. Being able to alter the goals of designs as well as the ability to understand their consequences is then of crucial importance.

2.4 Memory Management in Mobile Java

It is a general guideline that the programmer of a system running on top of a virtual machine should not attempt to optimize code based on the implementation of a virtual machine. The reason is that this can be error-prone, like any reliance on a particular implementation. Furthermore, a change in the implementation can change the rules of the optimization. Unfortunately, in mobile environments where memory is an important resource, also virtual machine and automated resource management become at least a partially leaking abstraction.

2.4.1 Motivation

As already discussed in Chapter 1, even when the underlying infrastructure is in principle managing resources automatically, it is important to consider data structures generated in a program. For instance, it is possible to harm garbage collection by using a data structure in an ill-minded fashion (Bloch, 2001). As an example, Bloch (2001) uses the following. Let us consider that we are implementing a stack of references in a language based on a virtual machine. The stack is implemented as a vector. An index (size) to the vector is used as the stack pointer in the abstraction as illustrated in Figure 2.4, where size, the actual vector, and some allocated

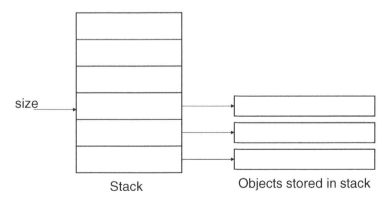

Figure 2.4   Stack of objects

items are included. When a new item is added to the stack, a new element in the vector is taken into use and the value of the stack pointer is incremented by one:

```
public void push(Object e) {
    ensureCapacity(); // Check that slots are available.
    elements[size++] = e;
}
```

Similarly, when an element is removed from the stack, the stack pointer is decremented by one. Whenever a new item is added, the new reference is added to the vector:

```
public Object pop() {
    if (size == 0) throw new EmptyStackException();
    return elements[--size];
}
```

Behaviorally, this program runs fine, and it passes all the functional tests when considering only input and output.

However, from the perspective of the garbage collector, it is unfortunate that there can be 'ghost' references to objects that are unusable from the viewpoint of the newly created stack abstraction, but remain valid and accessible in the vector used in the implementation. Such 'ghosts' are a result of executing several push operations followed by corresponding pop operations where the stack first grows and then gets smaller (Figure 2.5). The garbage collector is not allowed to deallocate them before the last reference to them has been erased, and having an accidental unused reference counts in this respect. In this case, the problem can be solved by setting vector element to zero in method pop:

```
public Object pop() {
    if (size == 0) throw new EmptyStackException();
```

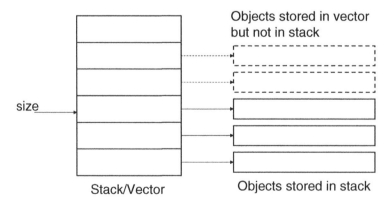

Figure 2.5 'Ghost' references in a stack of objects

```
      Object result = elements[--size];
      elements[size] = null;
      return result;
}
```

More generally, considering the behavior of the infrastructure forms an important part of composing programs even if the run-time infrastructure is principally handling all the resource management. Unfortunately, it is often hard to make appropriate design choices without knowing the details of the underlying implementation, and as a result, abstractions that do not internally behave as expected can be encountered. In the following, we give some mobile Java specific aspects that can be considered as design guidelines.

2.4.2 Rules of Thumb for Mobile Java

Downsizing memory usage depends on the design decisions of programmers. However, one should keep in mind that also other aspects than memory consumption should be taken into account, including topics like flexibility, maintainability, and performance. The principles discussed in the following have been introduced by Hartikainen (2005) and Hartikainen et al. (2006).

Avoid small classes. As classes introduce considerable overhead, it is better for memory consumption to merge small classes into bigger ones. In particular, inner classes that are easily created without further consideration are classes as well as regular classes. Usually inner classes contain only little functionality, like a particular listener of some user action for instance, so the overhead of having a class is relatively big in many cases. Using a lot of inner classes should therefore be deprecated. Finally, keeping the number of different exceptions as small as possible should be considered, because each exception is its own class with all the associated memory-related overhead, although this is not always too explicitly considered by developers. For instance, the memory consumption of an application that was implemented using two different structural alternatives, shrank to almost half from 14 019 bytes to 7467 bytes when the number of classes was reduced from 14 to 1 without altering the behavior of the application (Hartikainen et al. 2006).

Avoid dependencies. If the dependency between classes is not vital, one might save memory by removing references and using indirection via for instance an artificial identifier or some other class whose use cannot be circumvented instead. This saves at least an item from the constant pool of the class and in some cases the loading of the other class, which may not be an issue for a single instance but can be considerable in a large system.

Select the size when relevant, and manage vector/string usage. Vectors' library functions treat their contents as objects. Then, when something simpler is used as the content, the programmer can reduce memory consumption by using the native type. Moreover, whenever possible, by providing the size of the vector instead of using the default size one can reduce the memory footprint. Using strings may generate a lot of garbage, especially if strings are manipulated, as all manipulations generate new string instances. String buffer attacks this problem by reserving some extra space as a buffer so that a string can be expanded without having to make a new instance. This reduces the amount of generated garbage. If the size of a modified buffer is known beforehand, the buffer should be constructed with a defined `capacity`, or the `ensureCapacity` method should be called when using `StringBuffer` to ensure the correct size.

Consider using array versus using vector. Figures 2.6, 2.7, and 2.8 introduce three different implementations for storing a collection of integers. One implementation is based on an array, and the two others are based on using `Vector`. Table 2.1 shows the allocations made by each method when the value of `SIZE` is 2000 as introduced by Hartikainen (2005). Based on the figures, the difference between array- and vector-based implementations is relevant both in terms of the number of objects and used bytes. The methods that use vector effectively need to wrap integers to objects to be able to store their values, which explains most of the difference. Between vector implementations the difference in terms of created objects is minimal. On the other hand, an implementation where the correct size is given in the constructor, allocates up to 24% less memory in terms of size. This is because the other implementation (`vectorImplementationSimple`) has to increase the size of the vector in cases where the virtual machine has guessed the size of the

```
//
// Array based implementation.
//
private void arrayImplementation() {
    numberA = new int[SIZE];

    for(int i = 0; i < SIZE; i++) {
        numberA[i]= i;
    }
}
```

Figure 2.6   Array-based implementation

```
//
// Vector based implementation.
//
private void vectorImplementation() {
    numberV = new Vector(SIZE);

    for(int i = 0; i < SIZE; i++) {
        numberV.addElement(new Integer(i));
    }
}
```

Figure 2.7   Vector-based implementation

```
//
// Simple vector based implementation.
//
private void vectorImplementationSimple() {
    // Initialization with default size
    numberV2 = new Vector();

    for(int i = 0; i < SIZE; i++) {
        numberV2.addElement(new Integer(i));
    }
}
```

Figure 2.8   Simplified vector-based implementation using a default size

Table 2.1   Using strings vs. using StringBuffer

Method	Allocated bytes	Allocated objects
arrayImplementation	8 016	1
vectorImplementation	40 000	2 002
vectorImplementationSimple	52 000	2 010

```
//
// String based implementation.
//
public void useString() {
    String s = "";
    for(int i = 0; i < AMOUNT; i++) {
        s = s + "a";
    }
}
```

Figure 2.9   Using strings

vector in an erroneous fashion. The capacity of vector increases in increments, and therefore the difference in the number of created objects is not big.

Use StringBuffer. Concatenating String with the + operator or with the append method consumes memory as the virtual machine needs to create temporary objects. The results for using the procedures in Figures 2.9 and 2.10 are listed in Table 2.2 with the value of AMOUNT being set to 100 (Hartikainen, 2005).

Manage class and object structure. While enabling code reuse, inheritance can sometimes cost memory, as all the variables from the parent are present in a child object even though they might not be needed. When creating an object of a child class also its parent class needs to be loaded if it is not yet loaded in the system. The use of inheritance should therefore be carefully restricted to cases where it is necessary and useful. In particular, classes should not offer methods that are

```
//
// StringBuffer based implementation.
//
public void useStringBuffer() {
    String s = "";
    StringBuffer sb = new StringBuffer(AMOUNT);
    for(int i = 0; i < AMOUNT; i++) {
        sb = sb.append("a");
    }
    s = sb.toString();
}
```

Figure 2.10   Using StringBuffer

Table 2.2   Using strings vs. using StringBuffer

Method	Allocated bytes	Allocated objects
useString	39 000	450
useStringBuffer	304	5

not necessary. Alternate versions of methods can usually be avoided. In application design, choosing between offering a big class with a lot of methods or many smaller ones can also be difficult. In addition, cases where a hierarchy of classes is constructed can lead to superfluous loading and memory consumption.

Generate less garbage. Reusing old objects as already described above is one method to avoid making garbage.

Consider obfuscation. Since a majority of the content of a Java library often consists of metainformation and strings, a topic we will return to in Chapter 4, one can reduce footprint by obfuscating the names of public instance variables and methods, classes, and packages to a smaller form. Obviously, one should not obfuscate parts of systems that are visible to external parties. Moreover, standard facilities cannot be renamed for obvious reasons.

Handle array initialization. Long arrays static initializer can consume a lot of space as the Java compiler creates bytecode for static initializers of classes to initialize the array. Improved tool support can offer solutions that consume less memory. However, the savings will only be achieved with long arrays due to more complex routines. In practice, the difference only becomes meaningful when the size of an array is over 1000 (Hartikainen, 2005).

2.5   Symbian OS Memory Management

In contrast to infrastructure-managed memory handling used in mobile Java, in the Symbian OS environment it is the programmer who is responsible for allocating resources in an adequate fashion. However, unlike in C++ programming in general, the platform introduces a number of conventions for managing memory. While they can be considered as platform specifics, many of the conventions in fact aim at overcoming the known shortcomings of C++. In some ways, one can consider that the goal of the conventions is to guide the programmer to use abstractions in a fashion that prevents them from leaking, although in some cases this is only partial. In this section, we provide an overview of these conventions, together with the associated rationale.

2.5.1 Naming Conventions

Programming in C++ requires that memory is managed by the designer. The way Symbian OS is solving this problem is to introduce naming conventions that guide the developer to consider where variables are allocated, and to denote this in their names. The most important naming conventions are introduced in the following.

Class and Type Naming

- *Class names start with* C. Such classes are to be instantiated in the heap, and they should be derived from CBase either directly or indirectly for adequate memory management.
- *Kernel class names start with* D. The purpose of the convention is to separate kernel classes from application ones.
- *Mixin class names start with* M. The purpose of such classes is to introduce interfaces, and they are the only allowed form of multiple inheritance.
- *Type names start with* T. Such classes should effectively be used as a mechanism for reserving memory for a data structure, never for an object that requires actions in its destructor. Usually such classes are not allocated dynamically, but are either automatic variables or data members of other classes.
- *Enumerated types start with* E.
- *Resource names start with* R. This will ease treating them accordingly in code. In particular, any reserved resource must be released when it is no longer needed.

Method Naming

- *Method names start with a capital letter.*
- *Names of methods that can throw an exception end with* L. The purpose of the convention is to emphasize the use of exceptions as well as to help in reviewing completed programs.
- *Simple getters and setters reflect the name of the variable.*
- *Complex getters and setters are always to be given the form* GetSomeVariable *and* SetSomeVariable.

Variable Naming

- *Instance variable names begin with* i. The purpose is to highlight instance variables in programs. This highlighting becomes meaningful, as such variables should never enable memory garbaging. Instead, their allocation and deallocation is associated with the hosting object's creation and destruction.
- *Argument names begin with* a, which separates them from other variables in method code.
- *Constant names begin with* K. This is important, as constants can be allocated to ROM.
- *Automatic variable names begin with lower-case letters.* Preferably i and a are to be avoided in order not to risk mixing them with other variables.

Discussion

In practice, these conventions make Symbian programs easily identifiable as well as somewhat different from standard C++ programs. Moreover, also standard type names have been defined to reflect the conventions, which further eases recognition. Some sample names have been introduced in Table 2.3.

Following the above conventions is under the responsibility of the programmer in full, and errors in following the conventions will not prevent compilation if additional tools are not introduced. While this sometimes leads to hard-to-trace errors, it eases porting of other programs to the Symbian environment, assuming that only a shallow porting is performed, and no Symbian OS specific practices are introduced during its course. However, since supported libraries of C and C++ are not supported in full, it is possible that complications will occur even in a shallow embedding.

Finally, an error in following the conventions can be hard to track in a review, which is practically the only way to discover problems. The effect of such errors is considerable, as naming can also alter the way in which the different items should

Table 2.3 Sample Symbian names

Description	Example
Type	TInt
Instance variable	iIdentity
Constant	KMaxSize
Argument	aExpression
Enumerated type	EFalse
Kernel class	DSession
Method	LoadAddress
Resource	RThread
Namespace	NMyNameSpace

be treated by the programmer, in particular when allocating resources, which we will address later in this chapter.

2.5.2 Descriptors

Descriptors are the Symbian OS way to use memory in general, and they are also used in association with strings. Using descriptors takes place using dedicated descriptor classes. With strings, special macros denoting constant strings are used to enable their proper treatment in code generation. The following choices can be made by the programmer:

```
_L("Hello");
_LIT(KHelloRom, "Hello");
// String in program binary.

TBufC<5> HelloStack(KHelloRom);
// Data in thread stack.

HBufC* helloHeap = KHelloRom.AllocLC();
// Data in heap.
```

Of these formats _L is deprecated, and it should not be used except in test and debugging code, as explained by Stitchbury (2004). The reason is that the format has an overhead associated with constructing a run-time temporary TPtrC, which is a type that will be discussed in the following.

In addition to acting as containers of data, descriptors include a variable that represents the length of the string, enabling them to guard against overflows. Therefore, the following error would be noticed with a descriptor-based implementation but not with vanilla C++:

```
// Vanilla C++.
char userid[8];
strcpy(userid, "santa.claus@northpole.org");

// Symbian descriptor based implementation.
TBuf<8> userid;
_LIT(KSantasMail, "santa.claus@northpole.org")
userid = KSantasMail;
```

The result of the error would be a panic that terminates the execution of the thread.

The descriptor hierarchy introduced in the Symbian environment is illustrated in Figure 2.11. Of these, the following ones are probably the most commonly used in actual programs:

- TDesC is the base class for all descriptors. As indicated by C at the end of the name, which is yet another convention, the descriptor does not contain methods for manipulating it.

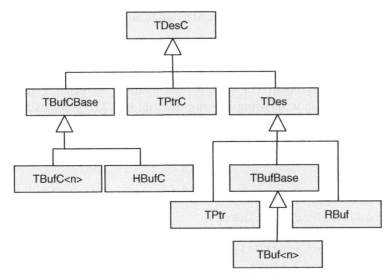

Figure 2.11   Descriptor hierarchy

- `TDes` is the base class for modifiable descriptors.
- `TBuf` and `TBufC` are descriptors that are to be allocated from stack.
- `HBufC` is the descriptor for allocating memory from the heap.
- `RBuf` is the modifiable version of `HBufC`. The type enables resizable descriptors, which are sometimes practical. The relation between the descriptors is similar to those between `TBuf` and `TBufC`.
- `TPtr` descriptors can be used to access (and edit) other descriptors. A constant version of this type of a descriptor is referred to `TPtrC`. These two descriptors are in fact references, which do not include a data buffer.

The layout of some commonly used descriptors is given in Figure 2.12. Considering the layout when composing programs is usually needed in order to perform transformations between different descriptor types.

For practical applicability of descriptors, the following guidelines are often adequate (Savikko, 2000):

1. Descriptors are commonly used instead of degenerating to using `TText*` format.
2. Form `const TDesC&` is used for parameters. This provides a light-weight (only reference is passed), safe (no accidental modifications), and simple (any type can be passed) solution.
3. Only instances of `HBufC` are allocated with `new`.
4. Conversions between different types of descriptors have been provided.

2.5.3   Exceptions

Symbian OS versions up to v.9.0 have relied on an in-house exception handling mechanism referred to as trap harness, where the basic operations are similar to the

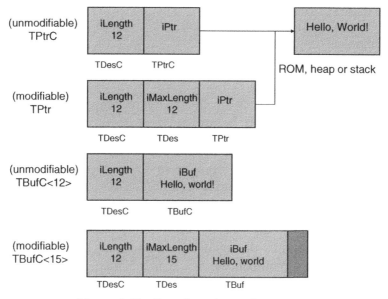

Figure 2.12 Descriptor layout in memory

standard, but have different syntax and slightly different semantics in the implementation. Even beyond v.9.0, the proprietary exception handling mechanism plays an important role as mitigating to the use of standard exception handling takes time. While the semantics of the handling mechanism enable a more light-weight implementation than standard C++ exceptions, the main elements are the same. A trap harness is used to denote the operation that is to be executed using macro TRAP (or TRAPD),[1] similarly to try in standard C++. Method User::Leave() corresponds to throw, and the actions to be taken when an exception is thrown (statement catch in standard C++) are usually implemented as a collection of conditional statements that define necessary recovery operations.

Let us next consider a simple example on using exceptions in Symbian OS. Assume that there is an enumerated type TMode with values EEat, ESleep, and EWait. The values of the type are treated differently in a function called BehaveL. This is achieved with the following definitions:

```
enum TMode {EEat, ESleep, EWait};

void BehaveL(TMode aMode)
    {
    switch (aMode)
        {
```

---

[1] The difference between the macros is whether the parameter used for identifying the potential exception has been declared or not.

```
    case EEat:
        // EEat specific behavior.
        break;
    case ESleep:
        // ESleep specific behavior.
        break;
    case EWait:
        // EWait specific behavior.
        break;
    default: // Beyond range of TMode
        User::Leave(KErrNotSupported);
        // Throws the exception.
        } // switch
    }
```

The above function is used by the following code snippet, where both trap harness and exception handling are introduced:

```
for (TInt i=0;; i++)
    {
    // Trap harness.
    TRAPD(error, BehaveL(static_cast<TMode>(i)));

    // Exception handling.
    if (error != KErrNone)
        {
        // Catch exceptions and recover from them.
        if (error == KErrNotSupported) Recover();
        } // if
    } // for
```

Notice that TRAPD (and TRAP) are normal macros and they bear no special meaning. The same holds for the routines recovering from the exception: they are straightforward C++ statements.

Finally, one should note that creating trap harnesses is a laborious operation from the performance point of view. Therefore, their number should be minimized whenever possible. However, sometimes it is a necessity to introduce a separate trap harness in a method that is not allowed to throw an exception, but which has to call another method that may potentially leave, even if in some sense exception handling seems superfluous. This is an unfortunate side-effect of the strict naming convention.

2.5.4 Combining Exceptions and Allocation

There is one more particular issue regarding the use of exceptions in the Symbian environment. A convention is that whenever memory is allocated from the heap, an overloaded version of operator `new` is used. This takes place in the following fashion, assuming that we are instantiating a new object of class `CMyClass`:

```
c = new (ELeave) CMyClass();
```

The semantics of the overloaded `new` operation are such that if everything is successful, a reference to the allocated object is returned. However, if memory runs out, the operation automatically throws an exception, thus eliminating the need to always check whether or not the allocation succeeded. Thus, one could replace all the places where this function is used with the following code snippet:

```
c = new CMyClass();
if (!c) User::Leave(KOutOfMemory);
return c;
```

As repeating this with all allocations in code would result in tedious programming tasks, overloading the `new` operator results in less code. An additional benefit is that programs are not polluted by low-level checks on whether or not memory reservation was successful, which releases the programmers to focus on application-specific issues.

When all the operations associated with a certain object have been completed, the object is deallocated. When performing this, resetting the value of the variable used is performed according to common C++ principles:

```
delete c;
c = 0;
```

While similar to other environments, this is more important in the mobile setting where applications can be turned on for an unlimited amount of time.

2.5.5   Cleanup Stack

A problem arises with the approach described above: what if references to objects allocated from the heap happened to reside in the execution stack in the areas that were removed due to the exception (Figure 2.13)?

In the Symbian OS environment, a special data structure has been defined, called cleanup stack, that should host all references that might be lost due to an exception (Figure 2.14).

The use of the cleanup stack is yet another task that is executed under the responsibility of the programmer. Whenever a new memory area (or object) is allocated from the heap so that the reference to it is an automatic variable that resides in the execution stack, the programmer must push the reference to the cleanup stack. When the object is deleted, it is again a task of the programmer to remove the reference from the stack. Pushing a reference to the stack is implemented with the operation:

```
CleanupStack::PushL(c);
```

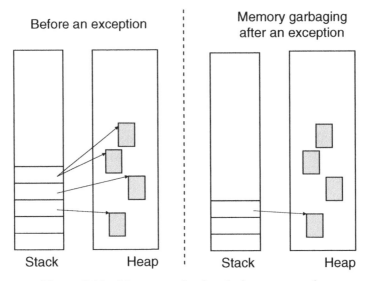

Figure 2.13   Memory garbaging during an exception

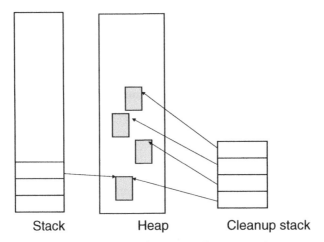

Figure 2.14   Using cleanup stack

where c is the variable being pushed. The corresponding pop operation is implemented as follows:

```
CleanupStack::Pop(); // c
```

Notice in particular the use of comments to denote the popped variable. This is important, as there is no generic tool support to help in checking that the right variable is popped, and reviews are often the only practical option to study the adequateness of the use of the cleanup stack.

If an exception takes place, the information in the cleanup stack is used to deallocate reserved resources. If the execution proceeds normally, the programmer should handle the removal of the reference from the cleanup stack, and optional deallocation and resetting the value of the variable according to common C++ principles.

There are several details related to the use of the cleanup stack. Firstly, in addition to the reference to the object, also a reference to a deallocating operation is included. As a result, one can control whether or not a destructor is executed when deallocating an object when an exception occurs. This is implemented with base class `CBase`, whose derivatives get their destructors called when cleaned from the cleanup stack during an exception, as `delete` is given as the method, making the called operation in fact `CleanupDeletePushL`. For other classes, no operation is defined, and as the result, only the allocated memory area the reference points to is deallocated. Moreover, due to the use of the reference to the deallocating operation, in addition to guarding against memory garbaging, the cleanup stack can be used for managing other resources as well. For instance, using `CleanupStack::ClosePushL(c)` defines `Close` as the deallocation method, and allows closing a session that is opened to a resource. Finally, one can also combine operations. For instance, to pop and delete an object with one code line one can use `CleanupStack::PopAndDestroy`, and give a parameter that defines how many items are popped from the cleanup stack (for example `CleanupStack::Pop(3)` that removes 3 items), which sometimes helps in writing readable code. Still, it is good practice to document what items are popped using comments, as otherwise reconstructing the contents of the stack for debugging purposes in reviews is made overly difficult.

A further special convention is introduced for denoting that a method adds something to the cleanup stack. Such method names should end with `LC` instead of plain `L` denoting the possibility of throwing an exception. Reflecting this with associated `Pop` is appreciated by other programmers who maintain the same code base. In contrast, `LD` denotes that something is removed from the cleanup stack, and the associated operation is executed when it has been defined.

An interesting design detail is that the cleanup stack itself can be safely used even if pushing an item to it may in fact cause an exception (hence the name `PushL`). The reason is that there always is enough room in the cleanup stack for at least one item. If the last slot was used, then more room is reserved, which may of course fail if the operation runs out of memory. However, at this point the reference to the new entry in the cleanup stack is already stored in the stack.

Finally, one should note that the use of the cleanup stack forces the designer to consider allocation and deallocation of objects. As references can only be removed from the top of the stack, deallocation of objects must follow the same guideline. This practically requires a design approach where allocations and deallocations are considered in an early phase of the development, and ad-hoc changes creating new objects are to be avoided. Furthermore, one should only use the cleanup stack for references to objects that might be lost when an exception occurs, never to objects that are parts of other objects in the form of instance variables, as this would at least potentially result in a situation where the same object would be deallocated several times (once via the cleanup stack and another time via the normal deallocation routine), at least potentially.

2.5.6   Two-Phase Construction

While the cleanup stack works well in most cases, there is one problematic situation. This is what to do if an exception takes place when a constructor is being executed. A reference to the object cannot be pushed to the cleanup stack, as it is not available before the completion of the execution of the constructor. On the other hand, resources may already have been reserved for the object, and releasing them after throwing the exception can become impossible, as there may be no handle to them. The following code example elaborates this:

```
C::C(int size)
    {
    iContents = new CVector(size);

    for (i = 0; i < size; i++)
        {
        iContents[i] = new (ELeave) CItem();
        }
    }
```

Now, if one is able to allocate memory for CVector and for some CItems but not for all of them, there is no way to deallocate the CItems that now exist; a thrown exception has removed all references to them, because the hosting CVector is destroyed. The situation is illustrated in Figure 2.15.

The Symbian OS defines a solution, where constructing objects is divided in to two phases, which are the normal constructor and an auxiliary method, usually named ConstructL. Then, operations that are necessary upon the construction of an object are partitioned so that statements that cannot cause an exception are located in the normal constructor, and all the statements that have the potential to cause an exception are put in ConstructL. In addition, a reference to the object to be constructed is pushed to the cleanup stack before calling ConstructL. At the level of program code, a programming idiom referred to as two-phase construction is used. We will address the idiom in the following.

When a new instance is created, the normal constructor is used, following the above fashion:

```
CData *id = new (ELeave) CData(256);
```

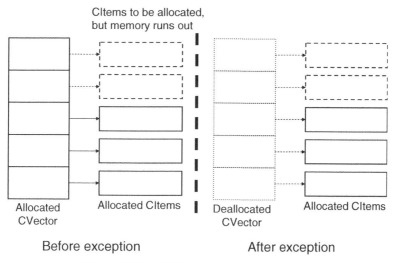

Figure 2.15 Failing constructor execution

This constructor can only leave when memory runs out, and in that case it should have no resources that have been reserved. There are no facilities that would ensure this, but it is a task of the programmer to design the constructor so that no exceptions can take place in its execution. Next, a reference to the reserved data structure is pushed to the cleanup stack, where it remains in a safe place in case of an exception:

```
CleanupStack::PushL(id);
```

following the normal Symbian practice. When a reference to the newly allocated object has been safely stored to the cleanup stack, we can call the second constructor of the object:

```
id->ConstructL();
```

which should be used for calling any operations that potentially may leave; hence the name `ConstructL`.

After `ConstructL` the actual operations that the object was created for can be executed. When all tasks have been performed, the reference should be removed from the cleanup stack in the normal fashion using `CleanupStack::Pop()`.

Special care must be given to cases where inheritance plays an important role. While standard constructors are executed automatically, executing the second-phase constructor for a parent class can require special measures. For instance, one can refer to the method with an explicit class name (form `CParent::ConstructL`) and with a special method name (for example `BaseConstructL`).

2.5.7   Factory Methods for Two-Phase Construction

As two-phase construction is a commonly needed procedure, auxiliary factory methods are often provided, which simplify the creation of certain collections of objects. In Symbian OS, such methods are often provided for hiding the two-phase construction. Factory methods are called `NewL` and `NewLC`, and they encapsulate the use of the cleanup stack and two-phase construction in object creation. Commonly used implementations are as follows:

```
CItem::NewL()
    {
    CItem * self = new (ELeave) CItem();
    CleanupStack::PushL(self);
    self->ConstructL();
    CleanupStack::Pop(); // self
    return self;
    }
```

```
CItem::NewLC()
    {
    CItem * self = new (ELeave) CItem();
    CleanupStack::PushL(self);
    self->ConstructL();
    return self;
    }
```

In practice, the implementation of `NewL` can be based on `NewLC` as well. This results in the following routine:

```
CItem::NewL()
    {
    CItem * self = CItem::NewLC();
    CleanupStack::Pop(); // self
    return self;
    }
```

Assuming that several objects are constructed, the routine generates less code in the binary output, and therefore is a commonly used implementation. Furthermore, it can be used without considering which second-phase constructor should be called.

Even if their use is commonly advocated, methods `NewL` and `NewLC` are not generated automatically, but the programmer should implement them whenever they are defined. Moreover, from the compilation or debugging perspective, these methods are just like any other user-defined methods, and they do not have any special characteristics. Their goal simply is to guide the developer using the code to correctly construct the object. Therefore, constructors and `ConstructL` can be introduced as protected member functions, which restricts the access to them from other classes but allows sophisticated construction by child classes.

2.5.8 Using Symbian Conventions

In this subsection, we give a sample Symbian program that can be compiled and run. The goal of the program is to demonstrate the use of the conventions as well as to provide a running Symbian example.

When composing a program we need to include Symbian conventions, including the type system, in it. This is performed by including some header files. In addition, we introduce a global variable `console`, which is used for communicating with the user. The resulting code is the following:

```
#include <e32base.h>
#include <e32cons.h>

LOCAL_D CConsoleBase* console; // Write messages to console.
```

The program contains a procedure similar to the one that was used as an example earlier. However, in this case, we do not use only functions but introduce a class that has a constructor, destructor, and a method that performs some sample processing and which is capable of leaving (`BehaveL`):

```
enum TMode {EEat, ESleep, EWait};

class CMyClass : public CBase
    {
public:
    CMyClass();
    virtual ~CMyClass();
    void BehaveL(TMode aMode);
    };

CMyClass::CMyClass()
    {
    console->Printf(_L("Constructor\n"));
    }
CMyClass::~CMyClass()
    {
    console->Printf(_L("Destructor\n"));
    }

void CMyClass::BehaveL(TMode aMode)
    {
    switch (aMode)
        {
    case EEat:
        console->Printf(_L("Eat\n"));
        break;
    case ESleep:
```

```
            console->Printf(_L("Sleep\n"));
            break;
        case EWait:
            console->Printf(_L("Wait\n"));
            break;
        default: // Beyond range.
            console->Printf(_L("Exception raising.\n"));
            User::Leave(KErrNotSupported);
            } // switch
    }
```

Next, we introduce a procedure that creates an instance of the above class and uses its services. To use the cleanup stack, we also introduce an automatic variable that will be used to store some dynamically allocated heap-based data:

```
void MyOperationL()
    {
    CMyClass * c = 0; // Automatic variable

    for(TInt i=0;; i++)
        {
        // Construction
        c = new (ELeave) CMyClass();
        CleanupStack::PushL(c);
        console->Printf(_L("Push.\n"));

        // Use
        c->BehaveL(static_cast<TMode>(i));

        // Destruction
        CleanupStack::Pop(); // c
        console->Printf(_L("Pop.\n"));
        delete c;
        c = 0;
        } // for
    }
```

Then, we need to give a function that will instantiate a console to which the above messages can be printed. In addition, the operation introduces a trap harness, which catches exceptions and recovers from them:

```
void MyConsoleL()
    {
    console = Console::NewL(_L("Cleanup example"),
                            TSize(KConsFullScreen,
                                  KConsFullScreen));
    TRAPD(error, MyOperationL()); // Trap harness
```

```
if(error != KErrNone)
    { // Exception handling.
    console->Printf(_L("Exception handled.\n"));
    } // if

console->Getch(); // Get and ignore character
delete console;
console = 0;
}
```

Finally, we introduce the main function `E32Main`, which will be the starting point of a Symbian executable. In addition, we introduce two auxiliary macros (`__UHEAP_MARK` and `__UHEAP_MARKEND`) that are used for managing the heap usage, constructing the cleanup stack, which would be handled by the Symbian OS GUI framework if we were implementing a real Symbian application but which must be created by the programmer in console applications, and calling the above operation using a trap harness. While such nesting is not really necessary in this program because no real operations are performed to recover from an exception, it is used to demonstrate the option in the following:

```
GLDEF_C TInt E32Main()
    {
    __UHEAP_MARK;
    // Get cleanup stack
    CTrapCleanup* cleanup=CTrapCleanup::New();

    TRAPD(error, MyConsoleL()); // Trap harness
    if (error!=TKErrNone)
        {
        User::Panic(_L("EPOC32EX"),error));
        }
    delete cleanup; // Destroy cleanup stack
    __UHEAP_MARKEND;
    return 0; // and return
    }
```

This completes the sample program, which is now ready for compilation and execution.

2.6 Summary

- Memory-related considerations are a necessity for implementing applications that will be run on a device that is constantly turned on.
- Preallocation and static reservation simplify memory management. Therefore, their use should be preferred especially when composing robust software.

- Linear data structures are preferable over their non-linear counterparts. The rationale is to reduce fragmentation as well as enable more efficient cache usage. Furthermore, it is sometimes possible to perform checks for the used range.
- Compression can be used to reduce memory consumption. However, its use can lead to increased processing of data, which in turn is commonly considered harmful in the mobile setting.
- Despite the use of patterns resulting in memory-effective code, composing maintainable programs requires balancing between contradicting requirements, some of which may be violated due to memory-effective design. Then, one should carefully consider a suitable compromise of different properties.
- Even if in the end a virtual machine, like in mobile Java, would be responsible for releasing unused resources, programmers' actions can considerably affect the resulting memory consumption.
- Design idioms and patterns are available for handling memory management. For instance, the Symbian environment defines a comprehensive set of conventions for managing memory in the C++ environment.

2.7   Exercises

1. What kind of data structure would be adequate for a calendar in a mobile device? How would adding of new items or deleting those that already exist be implemented? Could closing the application and opening it again change the data structure? What would such a dynamic data structure enable? How about the structure that is stored in the disk, assuming that the used physical implementation consists of flash memory?

2. How would you implement a data structure similar to the Symbian OS cleanup stack, but which would allow removal of items from the data structure based on name, reference, or some other identifier? How does your implementation correspond to the Symbian OS solution in terms of complexity and performance? Could the data structure be used for garbage collection as well?

3. What problems can you find in the following excerpt of a Symbian program? How and in what situations would they degenerate the program's execution? How should the program be corrected?

```
// Operations to be identified in the screen.
enum TMode {EEat, ESleep, EWait};

// Temporary resource for drawing.
class CMyResource
    {
public:
    CMyResource(CScreenShot & aParent);
    ConstructL();
```

```
    ~CMyResource();
    void ShowEatingItemL();
    void ShowSleepingItemL();
    void ShowWaitingItemL();
private:
    ...
    };

// Draw the right symbol to myWindow.
void Behave(TMode aMode, CScreenShot& myWindow)
    {
    CMyResource *res = 0; // Local drawing resource

    // Local two-phase construction.
    res = new CMyResource(myWindow);
    res->ConstructL();
    CleanupStack::PushL(res);

    // Selecting symbol to draw.
    switch (aMode)
        {
        case EEat:
            res->ShowEatingItemL();
            break;
        case ESleep:
            res->ShowSleepingItemL();
            break;
        case EWait:
            res->ShowWaitingItemL();
            break;
        default:
            User::Leave(KNoSuchActivity);
            break;
        } // switch

    // Drawing complete, resource no longer needed.
    CleanupStack::PopAndDestroy(); // res
    delete res;

    // Fix figure to the screen.
    MakeImagePermanentL(myWindow);
    }
```

4. Compose a procedure that calls itself recursively, and passes an object containing an array of 25 integers (each integer is 4 bytes, so the size is at least 100 bytes) as a parameter. How many rounds can the program execute in the Symbian

OS environment? Are there differences in the emulator and an actual Symbian device? What if the program is modified to reserve memory from the heap in a similar fashion? Does the allocation from stack or from heap affect performance?

5. What kind of design choices could be made to ensure that there are always enough resources for making an emergency call? Is reserving memory enough, or should there be also other means?

6. In what kinds of situations would it make sense to offer only `NewL` or `NewLC`, but not both, in the interface of a Symbian OS class? What are the fundamental differences in the use of these two methods?

7. What would be a realistic assumption for a third-party programmer on the amount of memory that will be made available for her program, assuming that only application-specific issues, not including any infrastructure requirements, are considered? How does the use of graphics, sounds, etc. affect the requirements of the application?

3

Applications

3.1 What Constitutes an Application?

In the conventional computing setting, programs can often be taken as transformators that translate a given input to a corresponding output. In practice, however, modern pieces of software take input and generate corresponding output in microscale, rapidly reacting to smallish changes in their environment like a touch on a touch screen, moving a cursor, or pressing a button. Moreover, changes can take place in parallel, i.e., rather than being sequential as traditional programs, applications have become reactive systems, which wait for any events that might affect them and potentially react to the events with some response. As events can arrive in different order or even in parallel, applications may become more complex entities, if they assume the responsibility for controlling the executions. Rather, the environment, for instance the user, associated network, or some other actor, more commonly takes control of what should take place.

The most basic definition of an application is that it is a piece of software that can be started and terminated individually, and that it performs a certain task. Furthermore, it is often necessary to associate a user interface with an application, as otherwise observing the behavior of the application might be difficult. In this chapter, we use the term 'application' in the broad sense, which includes the necessary user interfaces and related facilities, although the focus is placed on actual application software and common details of graphical user interface programming are overlooked.

In the technical sense, an application can be taken as a piece of executable code that can be triggered to execution by the user or the system under some special conditions. This, however, is not a necessary requirement but it is also possible to use an approach where one common executable loads all applications from a dynamic library, for instance, and calls a certain routine to activate the application using the factory method approach described above. In addition, in some cases, the application, say, a command shell, that starts other applications forks, i.e., creates

Programming Mobile Devices: An Introduction for Practitioners   Tommi Mikkonen
© 2007 John Wiley & Sons, Ltd

an identical replica of itself and uses the newly created copy for executing the other application.

In order to declare a piece of software an application, an interface mechanism must be defined, which tells the execution infrastructure that we are dealing with an application. The required definitions to glue an application to the infrastructure can be simple, or advocate a more complex definition of the application. For instance, in conventional C and C++ programming, such a definition is function `main`, from which the execution of the application begins. Usually, the more complex approaches also aim at guiding the developer to design using inheritance or in accordance to a certain design pattern, not just create an entry point for an application to begin its execution. In particular, in many cases platforms define a specific way to link applications to the surrounding infrastructure. A detail that has an effect on how the application is to be used is the depth of its integration to the rest of the system. At least the following cases can be identified:

- Application is independent of the rest of the system. Such types of applications are self-contained, and for the most part, they simply rely on the platform's low-level services. Implications for other applications can often be neglected.
- Applications share library code. In principle, this in fact takes place when the platform provides services to new applications. However, the more application-specific the library is, the more effort must be invested in keeping applications compatible with each other when the library version is altered.
- Application directly shares data with some other application. For instance, one can compose a personal information management and planning center application that collects all the data from contacts, todo list, calendar, and so forth. Then, when updates are made via one application, the data should be made available for other applications as well. As a result, applications are becoming tangled.
- Applications can be embedded in each other. For instance, when sending a multimedia message, it may be possible to run a camera application invisibly to the user to record the data to be sent.

The depth of integration is an obvious source of complexity. In particular, testing all the functions in different cases can be difficult.

3.2 Workflow for Application Development

Perhaps the most important design concern in the design of an application running in a mobile device is the consistency of user experience. This aspect can be affected by a number of design choices taken during the development of applications, resulting in consistent usability.

While the consistency of user experience is important, its design in the mobile setting is hard and requires taking users into account during design (Kangas and Kinnunen 2005). One principal problem is that in many cases users wish to perform rapid, focused actions, instead of long-lasting sessions, where users sometimes perform exploratory tasks to locate a certain feature, as can be the case with a desktop. Actions must be simple and single yet focused, and they must be accomplished with ease and using only a minimal number of keystrokes (Salmre 2005). This has an obvious effect on the way in which applications must be designed.

A common workflow for the development of applications for the mobile setting, with special focus on usability and user activities, has been defined by Salmre (2005), consisting of:

1. scoping,
2. performance considerations,
3. user interface design,
4. data model and memory concerns, and
5. communications and I/O.

In the following, we summarize this workflow.

3.2.1 Scoping

Before starting the design of an application for the mobile setting, one must have the fundamental purpose of the application, including both what the application can do and what it cannot. In particular, when implementing a mobile version of a desktop application, a subset of functions must be selected that will be included in the implementation. If needed, the features can be given relative importance, which allows the determination of the first set of functions. Furthermore, the physical characteristics of the device must be taken into account, if they imply restrictions.

Scoping can be helped by conceptualizing the application with pictures, mockups, and creating prototypes. This will also help when communicating the scope and the purpose of the application to other developers. One should also consider the relative importance of the functions to users. For instance, if clock times are rarely entered, it may be enough to use a somewhat inconvenient user interface; while the operation may be annoying, it is needed so seldom that the user can still execute it. However, for entries that are frequent, a well-considered user interface should be implemented.

3.2.2 Performance Considerations

When scoping has been completed, the next step is to consider performance. To begin with, general responsiveness metrics are needed for applications. This includes, for instance, defining how fast it should be to open a menu in the application. The overall responsiveness is an important part of the user experience. In addition to generic responsiveness, specific metrics should be created for the most important scenarios. This forces the application designer to consider the chains of events that allow the user to carry out certain procedures.

One way to design for performance is to use an older (or simply less capable) hardware for early experiments. While this gives a pessimistic view on the possibilities of implementing the application, the design can be initiated before the actual target device is available, and with lesser assumptions, it is more likely that the users will be satisfied with the performance.

All assumptions should be tested with a real implementation. A commonly used approach is to start with some key features and their performance, and to continue to less important features only when the key features have an acceptable level of performance. Taking into account that in the future more will be expected of the application is usually a good rule of thumb. In particular, an idea where the code is first completed in full in order to determine the worst bottlenecks is usually flawed, because the overall performance is often the most important aspect. Then, data structures, their layout in memory, used algorithms, and the way the user interface is constructed are issues that should be considered first, not individual lines of code. In other words, root causes of performance problems should be focused on instead of their symptoms.

One should also consider that overly focusing on performance can be harmful for portability. Therefore, while it is important to consider that the selected implementation principles are able to satisfy performance requirements, one should not be bound to optimize the development solely for performance. Rather, a reality check on what can be realistically accomplished is to be performed.

3.2.3 User Interface Design

As already discussed, before advancing to the technical design of a mobile application, it is important to study key use cases and features that characterize the application. If the performance provided by the prototype implementation is good enough in studies, it is time to focus on the right user interface.

Besides scoping and innovations, one can consider end-user productivity and responsiveness as the most important principles of user interface design. The former means that the actions that are typical and natural for the end-user can be easily and rapidly carried out. The latter means that the user has the feeling of being in control while performing the activities, which commonly implies minimizing the time the user has to wait for activities to complete, and even more importantly, the user is never left wondering what the device is actually doing. The design is further hardened by the tendency of users to perform repeated actions if no response is observed immediately. This encourages designs where feedback on user-initiated operations is given, even if the actual operation is still in progress behind the scenes. This may require a strategy where the user is tricked into believing that an already completed task takes place only on her command in a proactive fashion (for instance, some application can be always active even if the user has never started it), or that the device lets the user believe the task is completed while it in fact is not (for example, the phone claims to be ready after a reboot even if it has not yet loaded contacts from SIM). Moreover, in some cases one has to design an enforced flow of control, but at the same time avoid the user becoming frustrated. A further challenge is to keep the user aware of what has really been saved to disk, if the user wishes to turn off the device.

Of particular importance in designing the user interface are the available facilities. It is not realistic to copy the user interface greeted in one type of device to another type of device, and expect that usability and user experience will be preserved. Instead, one should consider what seems natural to the user when a certain type of device is available and use that as the starting point of user interface design. The situation is worsened by the fact that different actions are natural with different devices. For instance, it seems completely realistic to edit Excel macros when using a Communicator type of device, but being able to read the figures might be enough in a normal mobile phone where more restricted resources are available. In general, the design is of course influenced by the size of the screen as well as the restricted input mechanisms. To some extent, this can be solved by using PCs for some of the tasks, and only transferring the outcome to a mobile device.

In addition, one can consider whether to aim at special-purpose devices and applications or to a single tool that does everything. One view to this problem is provided by Norman (1998), where an application- and purpose-specific approach is considered to lead to simpler use than a multipurpose approach. In practice, however, it seems that also the latter approach is constantly gaining interest, at least when considering available devices. One contributing factor to this is the cost of manufacturing. New hardware features can be cheaper when they are integrated in a cellphone rather than implementing them in a separate device. Moreover, software features can be virtually free.

3.2.4   Data Model and Memory Concerns

As already discussed, mobile devices offer rather restricted facilities for application development. This is related to unit price of devices, where more sophisticated hardware leads to an increasing price per device, but also power consumption and the size of the device imply certain restrictions. The outcome can be a device where several handicaps exist, but the assumed use cases can be implemented with ease.

The way in which data is represented has an impact on how it can be located in the memory, on how the system behaves in peak conditions, and on how the application disposes data. For an application developer, this implies that data structures and memory use in general must be carefully considered. Also dynamically loaded libraries can be considered as an issue that is closely related to data model and memory concerns, as their technical implementation can rely on DLLs.

3.2.5   Communications and I/O

The way communications and I/O are defined determines how the application communicates with the resources that are located beyond its control. This includes devices' internal resources, such as files and subsystems, as well as resources that are external to the device, and require a communications mechanism before an access.

For instance, the latter includes socket-based communications, files on servers, Web Services, and remote databases, to name some options.

The way in which the application handles local and remote resources has a major effect on usability. Accessing local resources is usually fast, whereas communicating with remote resources is slow, at least with the current implementation techniques. A decision to load some data from a remote location in anticipation of the user's actions can in some cases result in major improvements in user experience. However, in general this is impossible, and should only be carried out in special cases, where users' intentions can be accurately modeled in advance.

Another important aspect to consider with communications and I/O is the level of abstraction of transmitted and stored data. For example, one can consider the following levels of abstraction in using files:

1. binary streams, where the data is stored in a fashion that is unreadable without auxiliary software,
2. text streams, where data becomes more readable, but may still remain somewhat unstructured and unreadable for a human reader,
3. XML forward-only readers and writers, where more meta-information is included,
4. XML Document Object Model, where complex automatic processing of included data is usually enabled.

The different levels of abstraction offer different facilities for manipulating data. The more abstract the level, the easier it is to process the data and the more self-contained the files are. This implies that developers' productivity improves, as programming, debugging, and maintenance will be easier, and it is more likely that potentially available standard components can be used, or reuse options exist within the company. However, at the same time the amount of overhead in transmitting, processing, and storing increases, which means that the approach may not be suited for cases where a large amount of data must be processed in a short period of time. This can lead to contradicting requirements in application development that complicate the design. The design is made more difficult by the fact that it is seldom a practical way to include several implementations of the same feature in the device, even if their characteristics would be different.

In addition, costs associated with the connection may become an important factor if a cellular data connection is assumed. For instance, one may wish to download as much data as possible when wireless LAN connection is available, but accept only minimal connectivity when using GPRS.

3.3 Techniques for Composing Applications

In this section, we introduce some implementation techniques that have been commonly used in the implementation of applications in a graphical user interface environment. These include event-based programming, the use of the model-view-controller (MVC) model, and the use of auxiliary files that can be used to describe attributes of applications, such as language settings and structural properties including menus, for instance.

3.3.1 Event-Based Programming

A commonly used approach to the programming of a graphical user interface is to allow every GUI element to generate events. For instance, when the cursor moves over a button, an event can be generated. Similarly, a button click can bear special significance only when executed over a certain graphical element depicting a button. This approach makes the graphical user interface an event generator, where events reflect the different choices of the user. Associating operations to the events provides a way to program such a system.

In general, callbacks are used to link events and code. Callbacks are special operations where one can register an operation that is called when a certain event, like a key press, occurs in an execution. The technique is commonly used in implementing event-based systems not only in the mobile setting but also in the workstation environment.

Implementing a callback can take many forms. One way is to pass a reference to a function during registration. When an event occurs, the function is called via the reference. An object-oriented implementation technique for event-based programming is to follow the Observer design pattern (Gamma et al. 1995). It allows registration of so-called observer objects to be interested in the changes in the state of a subject, which in turn commonly represents data. In the context of event-based programming, a common implementation is that events act as subjects, and when an application wishes to be notified about an event, it registers itself as the event's observer. Obviously, multiple observers can be registered to an event. Then, the occurrence of the event results in notifying all the observers.

Fundamentally, the difference between event-based programming and the traditional programming approach is that in the traditional approach, the service caller knows the provider of the service, whereas in the event-driven model, the event source only creates the event, and it is up to the components that have registered themselves to handle the event to perform their actions. The main benefits of event-based programming are simplicity and flexibility. The approach allows event handling to be serialized so that generated events are handled one by one. Usually, an approach is used where one event is handled in full before regarding the next event. This requires that programmers are prepared for events that can happen in any order, thus removing any causality that could be expected from the user interface. This in turn implies that the application must be solely based on event handling, and that any long-term executions must be executed in the background to avoid blocking of the user interface.

3.3.2   Model-View-Controller as the Application Architecture

In the following, we discuss the model-view-controller (MVC) design pattern. Introduced by Krasner and Pope (1988) the pattern has become a commonly used approach to systems relying on a graphical user interface. In addition to MVC, also the presentation-abstraction-control (PAC) pattern (Buschmann et al. 1996) is advocated by B'Far (2005) for the mobile environment. However, in many ways the patterns share the same ideas, and therefore only MVC is introduced in this context. Furthermore, also the Observer pattern is related to MVC, as in many ways one can consider a model as a subject whose state changes are notified to models and views. An event-based implementation can also be used, where the different events trigger operations into execution.

Following the ideas of separation of concern, the purpose of the MVC model is to separate all the functions of the application to one of the following three categories (Jaaksi 1995; Krasner and Pope 1988):

1. *Model*, which contains data structures of the application.
2. *Views*, which contain user interface(s) of the application.
3. *Controllers*, which allow the user to control the behavior of the application.

Figures 3.1 and 3.2 illustrate the structure of the model and operations associated with it. In the following, we discuss the different parts of the model in more detail.

Model. The role of the model in the MVC pattern is to host all the information regarding the data of the application. In addition, all operations regarding the data are included, following the guidelines of data encapsulation. The model is also usually responsible for managing data representation in permanent memory, as it can directly save and load data as needed. Whenever the data is manipulated, the

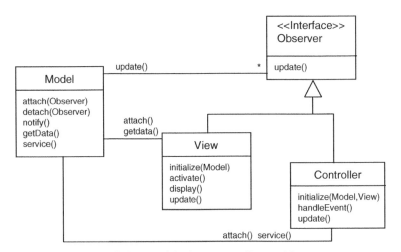

Figure 3.1   Model-view-controller architecture

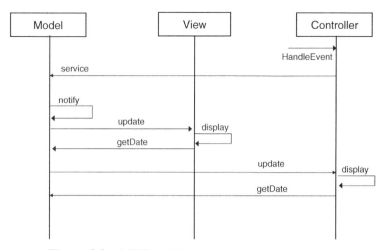

Figure 3.2 MVC architecture and dynamic behavior

model informs views and controllers about the change. As models remain relatively independent of all user interface operations, they can usually be reused in different environments.

View. Responsibilities of the view in the MVC design pattern are related to displaying the data to the user. Whenever a model is updated, the views that are registered to observe the model are notified about the update. Then, they query information about the update and display the upgraded data. A single application can contain multiple views for different purposes. This adds flexibility, as it is possible to select the view to use on the fly. A common problem in the use of the approach is that views can sometimes update themselves at an unnecessarily high frequency when a number of updates are made. While optimizations can be composed, the result is often a more complicated design. Views can sometimes be reused. A prerequisite for this is that the drawing operations are applicable also in the new setting. For a mobile device, this means that either a scaling mechanism is implemented, which may be harmful for performance, or reuse is restricted to devices whose screen resolution is compatible enough. Further restrictions result from properties of the screen, such as the number of colors or refresh frequency. Based on the above reasoning, the most common form of reuse in practice is probably the implementation of a collection of compatible view elements for a certain purpose, like statistics for instance. This enables the use of the same code in different applications. Furthermore, libraries can be implemented for certain domains that use similar facilities.

Controller. Responsibilities of the controller in the MVC model are related to controlling the application. The most common implementation is that a controller listens to the commands the application receives and translates them into a form that can be interpreted by the model. In a mobile setting, the option to reuse a controller is usually related to the configuration of the device that is available, because otherwise usability of the system can be degraded. For example, by creating a new configuration for the keyboard where number keys are in different locations, a previously intuitive control system can be ruined. Sometimes views and controllers can be integrated, if their operations are closely coupled. Integrating the components may make their reuse more difficult, as they usually become dependent on each other. Still, keeping the two concepts separate usually clarifies the structure of the application.

3.3.3 Auxiliary Files

On many occasions, applications may be relatively simple as such. However, when connected with graphics, their look and feel can improve considerably and in fact such data can form a considerable factor of memory consumption of a mobile application. Similarly, other aspects can also be separated from the actual application and presented in terms of so-called resources, where auxiliary data about the program can be stored. In the following, we discuss typical resources and other auxiliary files that are commonly needed by applications.

- *Menus*. In many environments, resource files can be used to define the structure of menus that are offered by applications. The rationale is that in order to modify menus, application code need not be modified, but it is enough to redefine some elements in the auxiliary file. Moreover, different language versions can benefit from this opportunity. In the simplest form, a menu is represented as a set of pairs that define a menu item and an operation associated with it, or a reference in one form or another. Then, the application concept is made responsible for initiating the execution of the correct operation, the identity of which can be passed as a parameter to some special method (for instance `HandleMenuCommand`). Obviously, by allowing additional menus as operations it is possible to create menu hierarchies.
- *Binary data*. Binary data is commonly needed in programs. For example, graphics, such as icons, are used for creating a special look for applications. While there are numerous formats that can be used in general, different platforms may have some restrictions on their use in different contexts. Connecting graphics to program behavior can take a similar form as with menus. However, icons can also bear significance on how an application is integrated to the rest of the system, enabling starting of an application using the icon. Similarly to graphics, other types of binary data can be used as well. For instance sounds and audio data are commonly used. Moreover, binary data could be used for security features.
- *Localization information*. Localization information is something that is commonly needed in applications. Such information includes definitions needed for multiple language versions. In some cases, they can be given together with other resources, but this is not a necessity. Instead, any file can be used for aiding in localization,

assuming that the rest of the program is composed respecting the localization principles. However, when the underlying platform offers its own practices for this, the platform's fashion should be predominantly used.

- *Other auxiliary files.* Auxiliary files needed by applications are many. For instance, there can be data files whose contents are needed for creating a valid initial state. Usually, one should compose applications so that even if some of the data files are missing, it would be possible to run the application. A related issue is that in some systems, it is advisable to implement applications such that they always have a default data file into which data can be saved if no filename is given for saving it. In addition to data files, there can be settings or profile files. Their goals can also be many, including for instance personal preferences or information regarding the properties of available resources. Again, it is commonly a good design goal to be able to run the application with minimal or even with fully missing settings files.

The importance of different auxiliaries is increasing. Firstly, facilities of mobile devices have improved, and as a result, it is possible to compose systems that benefit from better graphics and sound, for instance. Secondly, also expectations on wireless facilities have been increasing. Therefore, including data files for improved look and feel takes more memory than used to be the case.

3.3.4 Managing Applications

Like all applications, also applications in mobile devices need to be managed. This implies a number of operations as well as a format that is used to host applications before their installation. In this section, we address these issues.

A number of management operations can be associated with applications running in mobile devices. For instance, Riggs et al. (2001) introduce the following operations in the mobile Java environment:

- *Retrieval.* Applications can be retrieved from some location in the network or some other media. For instance, physical media such as memory cards can be used. However, for wireless devices, downloading applications over the air (OTA) is probably the most prominent alternative.
- *Installation.* Installs an application to a device. The installation may include several intermediate steps, such as verification that the installation is allowed, and transformation, where the downloaded software can be transformed to a suitable execution format.
- *Launching.* The application must obviously be startable once it has been downloaded and installed in the device.
- *Version management.* It may be necessary to upgrade already installed applications when new versions are released. It is also possible that upgrading one version of a subsystem (or application) requires an upgrade of other parts of the system as well.

- *Removal.* Applications can be removed from the device when they are no longer needed, or the storage space occupied by them is needed for some other use.

It is possible that applications that the user has downloaded are subjected to a different treatment than those that have been pre-installed by the manufacturer. For instance, the removal of the phone application is probably prohibited in every mobile phone, whereas uninstalling any user-installed piece of software is most likely expected. However, the manufacturer may also include application software in the phone that acts similarly to user-installed software. This software may introduce additional functions that have been implemented by a third-party company, and which are available for trials before an actual purchase. Moreover, they may sometimes be used for a restricted time before the payment.

In order to manage applications, it is common that they are delivered in different types of packages. The rationale for this approach is that while it is possible to copy files directly to a device's file system, like for instance in some Communicator types of devices, one should not expect that all users of mobile devices use a computer regularly. Instead, a user should be able to download a new game while traveling on a bus, resulting in a requirement to use over-the-air (OTA) download using the actual device only, not a development PC. Furthermore, it may be unacceptable to reveal all the files of an application, as this might induce problems with digital rights management, for instance. As a solution, most, if not all, mobile devices that can be extended with new software introduce also a format to be used for delivering the software. The contents of these formats can include executables, figures, auxiliary files including audio, graphics and video, and other resources. Packages can also include additional information that is used in the installation process as they can include information about supported platforms, versions that they rely on, and the size of the installation. Based on this information, it is possible for the installer to notice that the package is incompatible with a device. Moreover, the inclusion of version information can also be used for implementing a system where upgrades can be loaded on top of an already installed system, which makes the system more flexible. Similarly, information about uninstallation, licensing, payment possibilities, and so forth can be included. In addition to actual packaging, some environments require that the package is associated with a cryptographically created certificate. While this does not prevent the installation of potentially malicious software – the creator of the certificate can be misled, for instance – the certificate reveals the source the software originated from, which in turn can be used for stopping the application from further spreading.

Finally, assuming strict requirements with respect to installing only packaged applications to devices can lead to superfluous generation of installation packages at development time. In particular, applications that require some particularities of the device that cannot therefore be tested with the emulator only can force one to compose numerous installation packages as the development progresses.

3.3.5 Practicalities

It is common that different systems define their own ways to attach new applications to them. This is an immediate source of incompatibilities between different systems, because many mobile platforms require one to follow the established practices, and offer little room for variations for compatibility reasons. Still, the goal of the concept of an application is the same, to enable launching of new applications.

Another detail worth addressing is the complexity of the imposed application infrastructure. Some systems, like mobile Java, for instance, introduce a relatively simple application infrastructure that results in the eased construction of primitive applications. However, there is a cost related to the simplicity, as applications are often most naturally constructed so that while the model of the MVC pattern can be separated, the view and the controller are most conveniently integrated to avoid intimate interaction between two classes representing them. In contrast, in some other systems, like Symbian OS, for instance, a strict application architecture is imposed, where the main concepts remain the same, including all the components of MVC. The downside of this approach is that it can be overkill for small sample applications, making them seem overly complex. Furthermore, some systems that do not bear characteristics that could easily be associated with the infrastructure can become challenging to implement. For example, one can consider how a mobile Java virtual machine should be implemented following the MVC pattern; the problem is that the virtual machine simply does not play the role of an application but a piece of the execution infrastructure.

To summarize, rather than addressing any individual leaking abstraction as such in the application development, it seems that the different platforms reflect the different origins and intentions, which leads to different characteristics. As a result, they are suited for accomplishing different things. Moreover, the size of the step to compose the first application in a certain environment can vary considerably. For example, with mobile Java, as we will soon study, it is natural to start with only a single class and start extending the application. Symbian OS, on the other hand, shows signs of a system that has been targeted for more complex systems, where an established design has been composed before advancing to coding.

3.4   Application Models in Mobile Java

The main elements of a mobile Java execution environment, which are the most relevant for mobile Java, include the following elements:

1. *Configuration* defines the minimal requirements for the hardware of the device. The configuration also defines what kind of virtual machine is included in the system.
2. *Profile* defines the programming infrastructure available for applications intended to be run on top of a certain configuration.

3. *Additional interfaces and infrastructure* have been defined for accessing other facilities of the device. For instance, many resources can be accessed via such interfaces.

In the following, we discuss configurations and profiles separately. The most important commonality the different systems have is that their lifetime is determined by the infrastructure, not by the application models included in them, and therefore, we will pay special attention to this detail. However, also different application models are briefly introduced. At all levels, we focus on issues that help in dealing with restricted resources. Additional interfaces and resources they provide access to will be addressed later in Chapter 6. In general, many standard library interfaces have been removed to reduce memory footprint, and in some cases they have been replaced with simpler ones.

3.4.1   Configurations

As a configuration defines the minimal requirements for hardware, it can be taken as the contract between a hardware vendor and the developer of Java infrastructure. Therefore, it is the configuration that actually defines the type of virtual machine on top of which the system is implemented. In the context of mobile devices, the most important configurations are Connected Limited Device Configuration and Connected Device Configuration, which also set requirements for the underlying virtual machines. The relations between a hosting operating system, virtual machines, and configurations is illustrated in Figure 3.3.

Connected Limited Device Configuration

The simplest configuration applicable in mobile Java is Connected Limited Device Configuration (CLDC). It is intended to be run on top of a simplified virtual

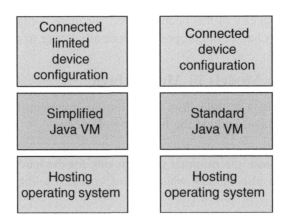

Figure 3.3   Mobile Java virtual machines and configurations

machine. Several implementations exist, with different features. For instance, the first CLDC-compatible virtual machine, Kilo Virtual Machine (KVM), includes design decisions taken for reduced memory footprint and simplified execution. For instance, stop-the-world garbage collection is used, and a design principle has been that on-the-fly compilation will be used conservatively only in cases where compilation can be hidden from the user, if at all. In contrast, CLDC Hotspot Virtual Machine implements the same basic features, but includes facilities for on-the-fly compilation, and thus allows improved application development. However, this is invisible to the programmer, because CLDC Hotspot Virtual Machine principally implements the same routines as KVM. The only difference is that the performance of some operations is improved when compiled code is benefited from.

Many, if not all, current CLDC virtual machine designs only support the use of one application per virtual machine. Therefore, in order to run several applications in parallel, one should instantiate one virtual machine per application, which is not a practical option from the memory consumption point of view. In fact, it is possible that devices are restricted to run only one Java application at a time to avoid loss of resources. However, design effort has been invested to implement a mobile device enabled virtual machine that would be able to run several Java applications in parallel. The design however requires some reconsideration of how security features of MIDP Java should work (Chapter 8).

As already mentioned, CLDC introduces some simplifications when compared to the normal Java environment to save memory, improve performance, and enable a simplified security concept that requires less memory and performance. The most important differences are the following:

- No support for floating point in CLDC v.1.0. The support was added in later versions of the standard. However, as a number of devices have no support for floating point operations, they often need to be implemented with software. This in turn can be slow and result in surprising performance problems.
- Simplified security scheme. The motivation for this is that the standard Java security scheme requires computationally complex executions. By simplifying the scheme, these executions can be simplified, which results in improved performance and reduced memory consumption. We will return to this topic in Chapter 8.
- No support for `finalize` operation. In general, using the operation in a mobile environment can be depreciated, because garbage collection can be implemented in a fashion that stops the execution of the application. Instead, one should use explicit deallocation that releases reserved resources.
- Thread groups or daemon threads are not supported.
- No user-defined class loaders are allowed. This is related to downsizing the code size of the virtual machine referred to earlier and to the simplified security scheme.

- No support for Java Native Interface (JNI). Only predefined access to device resources is offered, implying that all interfaces to the device are known at the time of installation. This simplifies the implementation of the virtual machine and security scheme. Moreover, memory footprint can be reduced.
- Weak references are only partially supported. No support is provided in the early version of the standard (v.1.0). Later versions include partial support.
- No features associated with reflection are provided. This enables smaller implementation of the Java system, as the support needed for this feature is a major source of memory consumption.

Many commercially available CLDC-based devices, low-end mobile phones in particular, can only offer a relatively small amount of memory for the Java environment, with the size of the Java environment being around 200 kb and individual applications around 40–50 kb. However, some high-end devices have these restrictions considerably relaxed.

Connected Device Configuration

In addition to CLDC targeted to low-end and middle-class mobile phones, mobile Java also comprises another configuration, called Connected Device Configuration (CDC). This configuration is based on the standard-featured Java virtual machine, and it assumes all the features of the full Java virtual machine to be available in the sense of the execution environment, but does not require all the libraries that are commonly used in the desktop environment.

Practical implementations can be optimized for mobile devices, and therefore they can be simplified in their implementation to conserve memory at the cost of performance, for instance. Therefore, while the standard-featured virtual machines in principle share the same structures, design decisions that define their performance and memory footprint vary. Still, in the design of a CDC-enabled virtual machine, the requirements for memory footprint and performance are increased considerably, when comparing it to CLDC virtual machines. As a result, the virtual machine can only be used in the most powerful mobile devices, such as PDAs and communicators, even if optimization regarding memory footprint and performance is carried out.

3.4.2 Profiles

While a configuration can be taken as a contract between hardware and Java environment vendors, a profile is a contract between the Java environment developers and application designers. Figure 3.4 depicts the main profiles of a mobile setting. In the following, we introduce the main characteristics of these profiles from the viewpoint of application models. For simplicity, we will however overlook details such as available interfaces and libraries etc. that are associated with them.

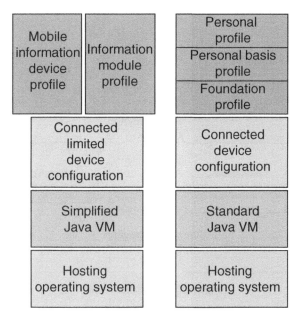

Figure 3.4 Mobile Java virtual machines, configurations, and profiles

Profiles Based on CLDC

Two standard profiles are defined for CLDC, referred to as Mobile Information Device Profile (MIDP) and Information Module Profile (IMP), which are in fact closely related. MIDP is a profile intended for devices that include a small screen, simple keyboard, and at least limited connectivity, proprietary mobile phones being the prime candidates for this profile. IMP, on the other hand, is derived from the MIDP specification by omitting features related to the screen. The purpose of this profile is to allow devices that can be operated from remote locations, including for instance different sensors that can be distributed to some research area, but operated remotely from a centralized facility. The main principles of MIDP Java have been introduced in an environment where mobile devices were still closed in the sense that only the device manufacturer could introduce additional features in its software. This resulted in a restricted environment, which was somewhat isolated from the functions of the phone. However, phones supporting MIDP Java have gradually adopted more and more features, allowing a more liberal access to the resources of the device.

Both MIDP and IMP define application models. MIDP applications are referred to as midlets and IMP applications as implets. As the application models are somewhat similar, we focus on the former in this representation.

Implementing a midlet is simple. One must derive the main class of the application from class `midlet`. In addition, methods must be provided for constructing, initializing (`startApp`), pausing (`pauseApp`), and destroying (`destroyApp`) the

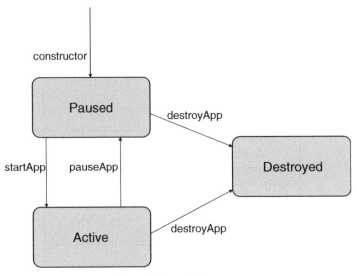

Figure 3.5   Midlet states

application. When a midlet is used, it makes transitions between these states. The internal state machine of a midlet is illustrated in Figure 3.5.

A generated and packaged application file, a so-called midlet suite, which was already addressed above, includes Java classes in an archived form, similarly to a normal JAR file, as well as all the associated auxiliary files, such as graphics or databases, that the application requires. Moreover, several midlets can be grouped into the same suite, assuming that they are allowed to share resources, which is not possible for midlets located in different suites. This is part of the security framework, which we will return to in more detail in Chapter 8.

In addition to actual midlets and resource files, one should include a manifest that defines details of the application. A MIDP manifest must contain, at the very least, the following fields:

- `MIDlet-Name`; the name of the midlet suite.
- `MIDlet-Version`; version of the midlet suite.
- `MIDlet-Vendor`; vendor of the midlet suite.
- `MIDlet-<n>`; names, icons, and midlet classes of the suite in an increasing order.
- `MicroEdition-Profile`; the profile that this midlet suite requires for installation and execution.
- `MicroEdition-Configuration`; the configuration this midlet suite requires for installation and execution.

Moreover, a number of additional manifest definitions can be introduced, giving more detailed information about the application, its download, and related aspects. For instance, the following definitions are available:

- `MIDlet-icon`; defines the icon used for the suite.
- `MIDlet-Info-URL`; a URL to additional information regarding the contents of the suite.
- `MIDlet-Jar-URL`; a URL to the JAR file containing the suite.
- `MIDlet-Jar-Size`; provides the size of the suite file. This can be used for determining whether or not it would be practical to download the application.

In addition to including the above information to the actual midlet suite, the same information should be given in a separate file also. This file, referred to as the Java Archive Descriptor (JAD), is helpful when deciding whether or not to download a particular file to a certain device.

The workflow for compiling JAR packages containing MIDP applications is somewhat different than in conventional Java or normal cross-compilation. To begin with, security features (Chapter 8) are supported by the practice of including additional information on any generated package. This additional information is generated by the development workstation, a new step in Java application development, and the goal is to ease the processing of an application when the application is loaded. In addition, the manifest described above must be included in the package.

Profiles Based on CDC

Similarly to CLDC, CDC also defines a collection of profiles that different devices can implement. Three profiles are commonly used, referred to as foundation profile, personal basis profile, and personal profile. While they are building on top of the standard-featured Java virtual machine, the facilities they offer for the application developer vary. In the following, we shortly characterize them:

1. The simplest profile, referred to as Foundation profile, mainly introduces basic classes of application development. However, no facilities associated with a graphical user interface have been provided.
2. Personal basis profile extends Foundation profile by offering the Xlet application model, which enables the development of simple applications. The states associated with the application model are illustrated in Figure 3.6.
3. The most complex standard profile building on CDC is Personal profile. In general, it resembles standard Java running in the workstation environment. It includes, for instance, the Applet application model (Figure 3.7) and a restricted version of Java bean technology. Moreover, it is possible to use the traditional application model, where the application takes control of its own lifetime (Figure 3.8). This can be beneficial when implementing applications that must be in execution constantly, such as those that for instance track the stock exchange or some other resource. In practice, the option to use an application model where all control is given to the developer results in application development not unlike that of Java application development in more conventional environments.

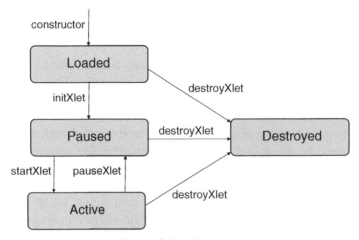

Figure 3.6 Xlet states

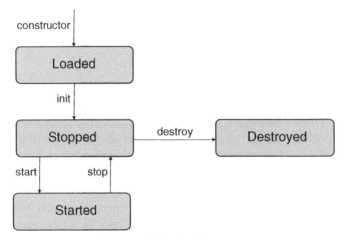

Figure 3.7 Applet states

Moreover, CDC on top of this profile offers an opportunity to use shutdown hooks that can be used to execute code when the virtual machine is shutting down.

As profiles based on CDC offer facilities that are fundamentally similar to standard Java, we will focus on the properties of CLDC-based profiles, in particular MIDP, which is targeted at mobile devices.

3.4.3 Sample MIDP Java Application

In this subsection, we introduce a sample midlet that displays simple statements and asks the user whether or not he agrees with them. Based on the answers, the

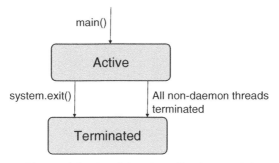

Figure 3.8 Traditional application model

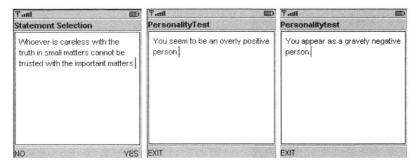

Figure 3.9 Sample Java application

system finally gives an evaluation of the personality of the user, as illustrated in Figure 3.9. In the following, we introduce the listing of the application, together with a discussion of its behavior.

To begin with, we import midlet and user interface libraries that will be used in this application:

```
import javax.microedition.midlet.*;
import javax.microedition.lcdui.*;
```

Like all midlets, the midlet we are implementing is derived from class `midlet`. In addition, the midlet implements the `CommandListener` interface for enabling commands from the mobile device's keyboard:

```
public class PersonalityTest
    extends MIDlet
    implements CommandListener {
```

Answers that the user can give to the system are received via `Command` objects (yes, no, exit). In addition, a `TextBox` is defined, which will be used for displaying the statements[1] to the user. This is defined as follows:

---

[1] Statements are from *Leadership Through the Ages. A Collection of Famous Quotations* published by Miramax Books, 2003.

```
private Command positive, negative, exitCommand;
/** Yes, No, Exit */

private TextBox tb;

/** Counter for statement and answer pair. */
private int nth, count = 0;

/** Questions */
private String Questions[] =
    { "The weapon of the brave is in his heart.",
      "A man is a lion for his own cause.",
      "Courage leads to the stars," +
      "fear toward death.",
      "Faced with crisis, the man of character " +
      "falls back upon himself.",
      "It is not the oath that makes us believe " +
      "the man, but the man the oath.",
      "Whoever is careless with the truth in " +
      "small matters cannot be trusted with "+
      "the important matters.",
      "No great thing is created suddenly."
    };

private int Qlen = Questions.length;
```

The constructor of the application creates three commands (exit from program, and the responses to a question) for receiving input from the user:

```
public PersonalityTest() {
    exitCommand = new Command("EXIT", Command.EXIT, 1);
    negative = new Command("NO", Command.CANCEL, 2);
    positive = new Command("YES", Command.OK, 2);
}
```

After the execution of the constructor, the midlet is in state idle. When the state of the midlet is set to run, method startApp() is called. In this particular midlet, the method picks a quotation that is shown to the user. In addition to startApp, all midlets are required to implement methods destroyApp and pauseApp, which are used to terminate and pause the execution of the midlet, respectively. These are given as follows:

```
protected void startApp() { pickStatement(); }

protected void destroyApp(boolean u) {}

protected void pauseApp() {}
```

The purpose of method `pickStatement` is to select one statement to be shown on the screen. Here, the selection is based on variable `nth`, which defines that questions are asked in sequence. This results in the following implementation:

```
private void pickStatement() {
    if (nth == Qlen) { giveInfo(); }
    else {
        displayStatement(Questions[nth]);
        nth = nth + 1;
    }
}
```

Display operation simply creates a new `TextBox`, attaches commands to it, sets this midlet as the listener of commands, and draws the `TextBox` to the screen. At the level of code, the following code lines are needed:

```
private void displayStatement(String statement) {
    tb = new TextBox(
        "Statement Selection", // Title.
        statement,             // Text.
        256,                   // MaxSize.
        0);                    // Constraints.
    tb.addCommand(positive);
    tb.addCommand(negative);
    tb.setCommandListener(this);

    Display.getDisplay(this).setCurrent(tb);
}
```

As already discussed above, this particular midlet allows the user to make selections. This has been implemented in method `commandAction`, which has been derived from interface `CommandListener`. In this case, the method records the answer and selects the next statement to be shown to the user:

```
public void commandAction(Command c, Displayable d) {
    if (c == exitCommand) {
        destroyApp(false);
        notifyDestroyed();
    } else {
        if (c == positive) { count++; }
    else {count--;}
    }
    pickStatement();
}
```

Finally, method `giveInfo` introduces a simple conditional expression that acts in accordance to the answers of the user (variable `count`):

```
    private void giveInfo() {
        if (count > 1) {
            tb = new TextBox(
                "PersonalityTest",
                "You seem to be an overly positive person.",
                50,
                0);
        } else if (count < -1) {
            tb = new TextBox(
                "Personalitytest",
                "You appear as a gravely negative person.",
                50,
                0);
        } else {
            tb = new TextBox(
                "Personalitytest",
                "You seem to have difficulties in being " +
                "in line with yourself.",
                80,
                0);
        }
        tb.addCommand(exitCommand);
        tb.setCommandListener(this);
        Display.getDisplay(this).setCurrent(tb);
    }
} // Closes class definition
```

In addition to the actual code, one must generate a Java Archive Descriptor (or midlet manifest) for this application. This task is supported with multiple tools. A file associated with this program is listed in Figure 3.10. With this auxiliary information, it is possible to generate an installation package, again using associated tool support.

3.5   Symbian OS Application Infrastructure

Symbian applications are commonly based on the MVC pattern enforced by the underlying application infrastructure. In addition, there are some auxiliary classes

```
MIDlet-1: PersonalityTest, PersonalityTest.png, PersonalityTest
MIDlet-Jar-Size: 1931
MIDlet-Jar-URL: PersonalityTest.jar
MIDlet-Name: PersonalityTest
MIDlet-Vendor: Unknown
MIDlet-Version: 1.0
MicroEdition-Configuration: CLDC-1.0
MicroEdition-Profile: MIDP-2.0
```

Figure 3.10   JAD file

Figure 3.11 Sample Symbian application

for connecting the pattern concept to the rest of the system, so that they can be initiated by the user.

In the following, we introduce a sample Symbian application, whose user interface is given in Figure 3.11. On the left-hand side of the figure, the application asks a question, in the middle, the user selects to answer the question, and on the right-hand side the application provides an answer. For a detailed discussion of Symbian OS application development, the reader is referred to e.g. Babin (2006).

3.5.1 Overview

When defining a Series 60 application,[2] the designer is assumed to define the following five different classes. Similarly to Java application models, life time of the application is again controlled by the infrastructure.

1. Application class, derived from CAknApplication, defines the concept of an application. The class also acts as the factory class for the application.
2. Document class of an application is defined from class CAknDocument. Instantiated by the corresponding application class, the document class defines the relation between the application's model and controller. Moreover, it can be used for storing the state of the application, when so desired.
3. UI class, derived from CAknAppUI, is used for defining the controller of the application. The class is instantiated by the corresponding document class. Another alternative is to use class CAknViewAppUI, which leads to a more controlled use of the MVC pattern.
4. View class is derived from CCoeControl. The purpose of this class is to define the view of the application. The view class is instantiated by the corresponding UI class. When using CAknViewAppUI the corresponding view base class is CAknView.
5. Engine class finally completes the MVC architecture by defining the model of the application. Engines are not obligatory, and especially when developing

---

[2] Series 60 applications include a platform-specific wrapping layer, denoted by *Akn* in class names. Plain Symbian applications use *Eik*, and other UI systems are used in an analogous fashion.

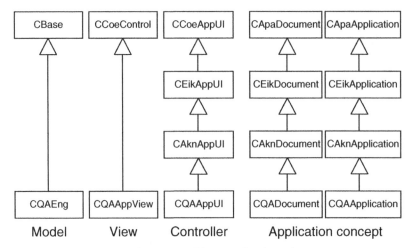

Figure 3.12   Symbian application classes

simple UI-intensive applications, their role can be neglected. However, for more complex cases, models can usually be reused in other types of Symbian devices, using some other user interface library. Engine classes are typically derived from `CBase`. Engines are usually instantiated by document classes, but also variations are possible.

In addition, a factory function `NewApplication` is provided to initiate the newly created application.

As an example, Figure 3.12 illustrates the structure of the sample application building on these classes. We will introduce the associated implementation in the following.

3.5.2   Resource File

In addition to actual code files, Symbian applications use a resource file that can be used for defining the details of the user interface of the application. To begin with, the resource file needs a name. Due to historical reasons, it can be four letters long. In addition, the resource file usually includes some headers that define resource structures and commonly used platform-dependent constants. In this particular case, we also include an additional auxiliary file, `qanda.hrh`, which contains application-specific constants to which we will return later. These items are listed as follows:

```
NAME qand

#include <eikon.rh>
#include <eikon.rsg>
#include <avkon.rh>
#include <avkon.rsg>

#include "qanda.hrh"
```

After NAME and include files, the following unnamed resource is commonly given:

```
RESOURCE RSS_SIGNATURE { }
```

The purpose of this resource is to enable the specification of version information.

Another resource that is to be defined is the name of the default file for saving the data associated with this application. As this application does not save its state to a file, the resource is not defined:

```
RESOURCE TBUF r_default_document_name { buf=""; }
```

Furthermore, a set of resources is given to define the menu structure of this application. This takes place at different levels of hierarchy. The following listing is used in the sample application:

```
RESOURCE EIK_APP_INFO
    {
    menubar = r_qa_menubar;
    cba = R_AVKON_SOFTKEYS_OPTIONS_EXIT;
    }

RESOURCE MENU_BAR r_qa_menubar
    {
    titles =
        {
        MENU_TITLE {menu_pane = r_qa_menu;}
        };
    }

RESOURCE MENU_PANE r_qa_menu
    {
    items =
        {
        MENU_ITEM { command=EQAAsk; txt="Ask"; },
        MENU_ITEM { command=EQAAnswer; txt="Answer"; },
        MENU_ITEM { command=EQAQuit; txt="Quit"; }
        };
    }
```

In addition to the resources and other items included in this sample file, it is possible to introduce strings in the resource file. This eases localization of applications, as the different language versions can be created more easily. For instance, when defining a descriptor in a resource file, one would give a definition:

```
RESOURCE TBUF32 r_qa_about_message
    { buf = "QandA by tjm@cs.tut.fi"; }
```

Then, when the string is to be used in a program, the following code snippet can be used:

```
HBufC* textResource =
          StringLoader::LoadLC(R_QA_ABOUT_MESSAGE);
```

This line would then assign the string to variable `textResource`. Notice `LC` in method `LoadLC`, which the programmer must take into account in the management of the cleanup stack.

Let us finally return to file `qanda.hrh`. A common way to link resource files to actual binaries requires that a constant value is defined for the variables that are used. In this application, there are four values that are used: `EQAAsk`, `EQAAnswer`, `EQAAbout`, and `EQAQuit`. They have been defined as follows:

```
enum TQAMenuCommands
    {
    EQAAsk = 0x1000,  // Own commands start from this value
    EQAAnswer,        // to avoid collisions with platform's
    EQAQuit           // values.
    };
```

Of the above, the different entries must be reconsidered for all applications. Moreover, if strings are included in this file, their contents must be defined per application. Therefore, it is not uncommon that the resource file gets relatively large in practical applications, especially in cases where deep menu hierarchies are used.

3.5.3 Attaching Application to Run-Time Infrastructure

In addition to defining the actual application classes, there are some other concerns. Firstly, Symbian application abstraction reveals the fact that they are actually dynamically loaded libraries (DLL, Chapter 4).[3] Secondly, as the Symbian infrastructure must be able to initiate the application, a special factory procedure must be given that initiates the application. In this particular case, the entry point and a procedure that initiates the application are given in file `qanda.cpp`.

The purpose of the entry point is to allow performing of special actions when the application is loaded but its execution is not yet started. For this application, there are no actions to be taken, and therefore the operation simply returns immediately:

```
#include "qandaapplication.h"
// Defines the class name of the application that is created.

// DLL entry point, return indication that everything is ok.
GLDEF_C TInt E32Dll(TDllReason /*aReason*/)
    {
    return KErrNone;
    }
```

---

[3] This only holds for Symbian versions preceding Symbian v.9. which has redefined some parts of the application infrastructure.

Finally, a factory operation is provided that creates a new instance of `CQAApplication`, and is called by the Symbian infrastructure when this application is started. This is given as follows:

```
// Create an application, and return a reference to it
EXPORT_C CApaApplication* NewApplication()
    {
    return (new CQAApplication);
    }
```

Newer versions of Symbian OS no longer implement application interface this way. Instead, plain `E32Main()` is used in the following fashion:

```
GLDEF_C TInt E32Main()
    {
    return EikStart::RunApplication( NewApplication );
    }
```

The reasons behind the modification lie in the implementation of security features that will be addressed in Chapter 8.

In summary, the characteristic property of the above code is the definition of the startup routine for the application. In doing so, it reveals the underlying implementation technique of the application either as a dynamically linked library (older Symbian versions) or an executable (newer Symbian versions).

3.5.4 Application

When developing Symbian applications in Series 60 environment, the base class that defines the concept of an application is `CAknApplication`. In the following, we give a listing for a sample application class.

Header

In the header part of the application class, only a few items are needed. These are the definition of a method for querying the application's unique identifier UID. Values for actual applications need to be obtained from an official source. However, depending on the Symbian OS version, some ranges are available for development and testing purposes. Prior to Symbian OS version 9.0 (S60 v. 3.0), range `0x01000000-0x0FFFFFFF` is to be used, whereas after that range `0xE0000000-0xEFFFFFFF` is available for free use. In addition, we introduce a method for creating the document, which we will discuss in the following subsection. These are introduced as follows in file `qandaapplication.h`:

```
#include <aknapp.h> // Details of application base class.
class CQAApplication : public CAknApplication
    {
public:  // from CApaApplication
```

```
    TUid AppDllUid() const;
protected: // from CEikApplication
    CApaDocument* CreateDocumentL();
    };
```

The included header file `aknapp.h` is needed for being able to use the concept of application in Series 60 environment. Method `AppDllUid` is for querying the identity of the application, and method `CreateDocumentL` is for instantiating the document class associated with this application.

Implementation

In the following, the above header is implemented (file `qandaapplication.cpp`). The different parts of the implementation have the following purposes.

First, header files are included for defining the concept of application and document, which will be created by the application. In addition, a constant is defined (`KUidQAApp`), which defines the UID of the application. This results in the following definitions:

```
#include "qandadocument.h"    // Document creation.
#include "qandaapplication.h" // Own header.

// Local constants.
static const TUid KUidQAApp = {0x01005b97};
```

Second, an implementation is provided for document creation. This operation simply makes a method call to a constructor of the document class associated with this application. The operation is defined as follows:

```
CApaDocument* CQAApplication::CreateDocumentL()
    {
    // Create QA document, and return a pointer to it
    CApaDocument* document = CQADocument::NewL(*this);
    return document;
    }
```

Finally, an operation is provided for querying about the application-specific identifier of this particular application. This is a standard Symbian feature, which can be implemented as follows:

```
TUid CQAApplication::AppDllUid() const
    {
    return KUidQAApp; // Return the UID
    }
```

This completes the methods needed for class `CQAApplication`, and the introduction of application-related facilities. In the following, we move on to the concept of document that defines the content of applications.

The main contribution of this class is the introduction of the application's identity on one hand, and the delegation of the application's creation on the other hand. Of these, the former is obviously unique, whereas the latter can be similar in many applications.

3.5.5 Document

As already hinted, a document is an important concept in Symbian application architecture. Created by applications as discussed above, the purpose of documents is to allow applications that follow the MVC model discussed earlier in this chapter. In other words, while the application class is responsible for implementing the concept of an application, documents are commonly used for maintaining the relationship between user interface independent models (or engines, in accordance with Symbian terminology) and user interfaces.

Header

First, we again include a number of header files. They are needed to be able to use the concept of document as defined in the Symbian environment. In addition, the document addresses the engine of this application and user interface. They will be defined in files `qandaappui.h` and `qaeng.h`, respectively. This results in the following code snippet:

```
#include <akndoc.h>       // Document base class.
#include "qandaappui.h" // Controller that will be created.
#include "qaeng.h"       // Model that will be created.
```

Whenever a document is introduced in Series 60 environment, the class from which the document is derived is `CAknDocument`. The definition includes standard methods for construction and destruction of documents, a method for creating a user interface (`CreateAppUiL`), and a method that gives a reference to the current engine of the application (`Model`), which also is the only instance variable we introduce in the document. This results in the following definition:

```
class CQADocument : public CAknDocument
    {
public:
    static CQADocument* NewL(CEikApplication& aApp);
    static CQADocument* NewLC(CEikApplication& aApp);
    ~CQADocument();
    CEikAppUi* CreateAppUiL();
    CQAEng* Model();

private:
    void ConstructL();
    CQADocument(CEikApplication& aApp);
```

```
CQAEng * iModel;
};
```

In addition to the above operations, the document class can be used for automatic loading and saving of application state (Externalize and Internalize). In this application, however, we will omit such details of Symbian application development.

Implementation

The implementation of the document class, given in file qandadocument.cpp, begins by including the necessary header files, listed in the following:

```
#include "qandadocument.h" // Own header.
```

Next, we introduce the standard Symbian OS construction sequence, which in this case consists of methods NewL, NewLC, ConstructL, and normal constructor. Following the principles of two-phase construction, all code that might lead to an exception is located in method ConstructL. In this particular application, such an operation is the creation of the model, which is included in the listing. Furthermore, a destructor is given that destroys the model. These are listed as follows:

```
CQADocument* CQADocument::NewL(CEikApplication& aApp)
    {
    CQADocument* self = NewLC(aApp);
    CleanupStack::Pop(self);
    return self;
    }

CQADocument* CQADocument::NewLC(CEikApplication& aApp)
    {
    CQADocument* self = new (ELeave) CQADocument(aApp);
    CleanupStack::PushL(self);
    self->ConstructL();
    return self;
    }

void CQADocument::ConstructL()
    {
    iModel = CQAEng::NewL();
    }

CQADocument::CQADocument(CEikApplication& aApp)
    : CAknDocument(aApp)
    {
    // Add any construction that cannot leave here.
    }
```

```
CQADocument::~CQADocument()
    {
    if (iModel) delete iModel;
    // Add any destruction code here.
    }
```

In addition to construction and destruction, two more operations are introduced. First, method `CreateAppUiL` constructs a user interface for this particular application. This is given as follows:

```
CEikAppUi* CQADocument::CreateAppUiL()
    {
    // Create application user interface and return a
    // pointer to it.
    CEikAppUi* appUi = new (ELeave) CQAAppUi;
    return appUi;
    }
```

Second, an operation is provided for requesting the model that is stored inside this document. This simply returns the associated instance variable, resulting in the following implementation:

```
CQAEng * CQADocument::Model()
    {
    return iModel;
    }
```

This completes the definition of the document, where the most common application-specific features are related to the creation of the application. Next, we will set the focus on defining the actual user interface of the application.

3.5.6 User Interface

The user interface of this application is implemented in class `CQAAppView`, which is defined in files `qandaappui.h` and `qandaappui.cpp`. The contents of the files implement the controller part of the MVC model. The files are listed in the following.

Header

The most important aspects of the header file are the facts that the application UI is derived from class `CAknAppUi`, and that we override command handler method `HandleCommandL`, which gets the commands defined in file `qanda.hrh`. In addition, the controller knows the view and the model that are active (`iAppView` and `iModel`). The contents of the file are the following:

```
#include <aknappui.h> // Generic application UI.
#include "qaeng.h"     // Model.

class CQAAppView; // Forward declaration for the view.
class CQAAppUi : public CAknAppUi
    {
public:
    void ConstructL();
    CQAAppUi();
    ~CQAAppUi();

public: // from CEikAppUi
    void HandleCommandL(TInt aCommand);

private:
    CQAAppView* iAppView;
    CQAEng* iModel;
    };
```

Implementation

We use a number of additional header files in order to use the the items defined in
the resource file (compiled to qanda.rsg), as well as the different constants given
in different files. The actual headers we include in the application are the following:

```
#include <avkon.hrh>    // Standard constants.
#include <qanda.rsg>    // Stuff generated from resource file.
#include "qanda.hrh"    // Application-specific UI constants.

#include "qandadocument.h" // Document that will be accessed.
#include "qandaappui.h"    // Own header.
#include "qandaappview.h"  // View that will be accessed.
```

The contents of the controller are simple. Again, the majority of the code lines
are used for defining the normal Symbian constructors and the destructor. They are
listed in the following:

```
void CQAAppUi::ConstructL() // Called by application framework
    {
    BaseConstructL();
    iModel = static_cast<CQADocument*>(iDocument) -> Model();
    iAppView = CQAAppView::NewL(ClientRect());
    iAppView->SetModel(iModel);
    AddToStackL(iAppView); // Series 60 practice; set the view
                           // active to receive key events.
                           // When using CAknViewAppUi, the
                           // corresponding method is AddViewL.
```

```
    }
CQAAppUi::CQAAppUi()
    {
    // add any construction that cannot leave here
    }

CQAAppUi::~CQAAppUi()
    {
    if (iAppView)
        {
        RemoveFromStack(iAppView);
        delete iAppView;
        iAppView = NULL;
        }
    }
```

The actual contribution of this class to the application is defining operations for the different commands that the user can give as input. In this application, all the input must be selected from a menu, and all the different selections have separate operations, identified by the values defined in file qanda.hrh and standard headers (EAknSoftkeyExit). If an unknown command is selected, panic is raised, which terminates the thread running the application. The method has been implemented as follows:

```
void CQAAppUi::HandleCommandL(TInt aCommand)
    {
    switch(aCommand)
        {
    case EQAAsk:
        iModel->Reset();
        iAppView->DrawNow();
        break;
    case EQAAnswer:
        iModel->Select();
        iAppView->DrawNow();
        break;
    case EAknSoftkeyExit:
    case EQAQuit:
        Exit();
        break;
    default:
        User::Panic (_L("QandA"), KErrNotSupported);
        break;
        }
    }
```

   This completes the introduction of the controller, where the application specifics
are most visibly included in the interpretation on how to handle different user
interactions. For obvious reasons, this is something that requires explicit design
attention. Moreover, the design is related to the resource file, as they use the same
variables for communication between the definition of the menu hierarchy and actual
effects in code.

3.5.7 View

The view of this application directly corresponds to the view of the MVC model.
The class is responsible for drawing the selected question and the selected answer
to the screen when appropriate. Timing of drawing them to the screen is directly
associated with user activities, i.e., when the user selects a new question or an
answer, the screen is redrawn. Defining what to draw is fully handled by the engine,
which is only addressed by this class.

Header

Unlike other classes used in application development, views are not derived from
a related base class whose name would somehow link to view. Instead, views are
derived from so-called control elements, in this particular case from CCoeControl
base class. Operations are given for construction, destruction, setting the model
based on which drawing is performed, and the actual draw operation. The following
listing is given in file qandaappview.h:

```
#include <coecntrl.h> // Generic controls.
#include "qaeng.h"    // Model.

class CQAAppView : public CCoeControl
    {
public:
    static CQAAppView* NewL(const TRect& aRect);
    static CQAAppView* NewLC(const TRect& aRect);
     ~CQAAppView();
    void SetModel(CQAEng * aModel);

public:  // from CCoeControl
    void Draw(const TRect& aRect) const;

private:
    void ConstructL(const TRect& aRect);
    CQAAppView();
    CQAEng * iModel;
    };
```

   Next, we address the associated implementation.

Implementation

The implementation of methods of class `CQAAppView` is given in file `qandaapp-view.cpp`. The file includes the above header as well as some additional files for using the items described in the resource file as well as to use the contents of controls, listed in the following:

```
#include <coemain.h>      // Control related.
#include <qanda.rsg>      // Compiled resource file.
#include "qandaappview.h" // Own header.
```

Construction and destruction take place similarly to the common routine as already discussed in several other classes, listed as follows:

```
CQAAppView* CQAAppView::NewL(const TRect& aRect)
    {
    CQAAppView* self = CQAAppView::NewLC(aRect);
    CleanupStack::Pop(self);
    return self;
    }


CQAAppView* CQAAppView::NewLC(const TRect& aRect)
    {
    CQAAppView* self = new (ELeave) CQAAppView;
    CleanupStack::PushL(self);
    self->ConstructL(aRect);
    return self;
    }

void CQAAppView::ConstructL(const TRect& aRect)
    {
    // Create a window for this application view.
    CreateWindowL();
    SetRect(aRect); // Set the window size.
    ActivateL(); // Activate the view.
    }

CQAAppView::CQAAppView()
    {
    // Add any construction code that cannot leave here.
    }


CQAAppView::~CQAAppView()
    {
    // Add any destruction code here.
    }
```

Setting the model simply sets the instance variable to the parameter of the method:

```
void CQAAppView::SetModel(CQAEng * aModel)
    {
    // Document owns the model, reference used here.
    iModel = aModel;
    }
```

Finally, method `Draw` is responsible for drawing the applications view to the screen. Drawing is based on asking questions and answers from the model of the application, available to the view by using instance variable `iModel`. The method is listed in the following:

```
void CQAAppView::Draw(const TRect& /*aRect*/) const
    {
    // Get the standard graphics context.
    CWindowGc& gc = SystemGc();

    TRect rect = Rect(); // Gets the control's extent.
    gc.Clear( rect ); // Clears the screen.

    gc.UseFont( iCoeEnv->NormalFont() );
    gc.DrawText( iModel->Question(), TPoint(5,30) );

        if ( iModel->Used() ) {
            gc.DrawText( iModel->Answer(), TPoint(5,60) );
        }
    }
```

This completes the introduction of facilities needed for a graphical user interface. For obvious reasons, the most important things to define in an application-specific fashion are related to drawing to the screen. Finally, we can now define the model that the above classes use when addressing data structures.

3.5.8 Engine

In this particular application, the engine is implemented in a class called `CQAEng`. It manages all questions and answers that can be provided for them, and has been given in files `qaeng.h` and `qaeng.cpp`.

Header

In the technical sense, questions and answers are both implemented as arrays of descriptors, named `iQuestions` and `iAnswers`, respectively. These arrays are initialized when constructing the engine. Operations are provided for construction and destruction of the engine, for accessing both questions and answers, for selecting a new answer, for checking if the current question−answer pair has already been used, and for resetting the system. The above design results in the following header file for the engine:

```
#include <e32std.h>   // Standard Symbian stuff.
#include <e32base.h> // Standard Symbian stuff.
#include <badesca.h> // Array usage from badesca.h.

class CQAEng : public CBase
   {
public:
   static CQAEng* NewL(); // Two-phase constructor.
   ~CQAEng(); // Destructor.

   const TPtrC Question(); // Reference to current question.
   const TPtrC Answer(); // Reference to current answer.

   void Select(); // Allows one to answer a question.
   TBool Used(); // Identifies whether the answer is visible.
   void Reset(); // Defines new question and answer.

protected:
   void ConstructL(); // Constructors.
   CQAEng();
private:

   // All questions and answers implemented as arrays of
   // descriptors. Also class RArray could be used.
   CDesCArrayFlat * iQuestions, * iAnswers;

   // Current question and answer are indices to the arrays.
   TInt iQuestion, iAnswer;
   TBool iUsed; // Identifies if the answer has been selected.
};
```

Notice the introduction of constructors as protected. The goal is to prevent their accidental use instead of factory methods that implement two-phase construction.

Implementation

The above methods have been implemented as follows. First, standard and engine-specific header files are included, and the standard two-phase constructor is introduced. In addition, the numbers of questions and answers that will be used are defined in this file:

```
#include <e32std.h> // Standard Symbian stuff.

#include "qaeng.h" // Own header.
```

```
const TInt KQuestionCount = 4;
const TInt KAnswerCount = 5;

CQAEng * CQAEng::NewL()
    {
    CQAEng* self = new (ELeave) CQAEng;
    CleanupStack::PushL(self);
    self->ConstructL();
    CleanupStack::Pop();
    return self;
    }
```

The methods called by NewL are responsible for resetting the values of instance variables to initial values (normal constructor), and for initializing the arrays of questions and answers (ConstructL) based on the given strings. The partitioning of operations in the two different constructors is directly based on whether or not they potentially throw an exception, following the normal Symbian two-phase construction scheme. This results in the following code:[4]

```
CQAEng::CQAEng() // Resets instance variables.
    {
    iQuestion = 0; // strictly speaking not necessary, because
    iAnswer = 0;   // 0 is the default value.
    iUsed = EFalse;
    }
void CQAEng::ConstructL() // Creates questions and answers.
    {
    iQuestions = new (ELeave) CDesCArrayFlat( KQuestionCount );
    iQuestions->AppendL( _L( "Can I stay up?" ) );
    iQuestions->AppendL( _L( "Let's watch TV?" ) );
    iQuestions->AppendL( _L( "Eat doughnuts?" ) );
    iQuestions->AppendL( _L( "Let's go swimming?" ) );

    iAnswers = new (ELeave) CDesCArrayFlat( KAnswerCount );
    iAnswers->AppendL( _L("Not me.") );
    iAnswers->AppendL( _L("Not my problem") );
    iAnswers->AppendL( _L("Ask my brother.") );
    iAnswers->AppendL( _L("Obviously.") );
    iAnswers->AppendL( _L("Not interested.") );

    Reset(); // Pick the first question and answer.
    }
```

---

[4] Individual questions and answers are based on numerous conversations with the author's children.

Obviously, the mechanism of using resource files for defining strings would be an improvement over this approach. However, for simplicity, we continue with this implementation in the example.

When the application terminates, memory reserved for questions and answers is released, provided that the allocation has taken place in the first place. This has been handled in the destructor as follows:

```
CQAEng::~CQAEng()
    {
    if (iQuestions) delete iQuestions;
    if (iAnswers) delete iAnswers;
    }
```

A number of routines are offered for the user interface. First, the user interface can get the current question and answer from the engine (methods Question and Answer). Second, the user interface can select an answer from the menu, in which case the engine is notified (Select). Finally, the user interface can query whether or not the current question has been answered (Used). All these methods have straightforward implementations listed in the following:

```
const TPtrC CQAEng::Question()
    {
    return (* iQuestions)[iQuestion];
    }
const TPtrC CQAEng::Answer()
    {
    return (* iAnswers)[iAnswer];
    }
void CQAEng::Select()
    {
    iUsed = ETrue;
    }
TBool CQAEng::Used()
    {
    return iUsed;
    }
```

Finally, Reset selects the question and the answer from the arrays in an exhaustive fashion, and resets the question–answer pair so that only the question will be drawn to the screen when the screen is redrawn the next time:

```
void CQAEng::Reset()
    {
    iQuestion = (++iQuestion) % KQuestionCount;
    iAnswer = (++iAnswer) % KAnswerCount;

    iUsed = EFalse;
    }
```

```
; qanda-simple.pkg
;

;Language; standard language definitions
&EN

; standard SIS file header; application name and id.
#{"qanda-simple"},(0x01005b97),1,0,0

;Supports Series 60 v 0.9
(0x101F6F88), 0, 0, 0, {"Series60ProductID"}

; Files to copy; Application and compiled resource file.
"..\..\..\epoc32\release\armi\urel\qanda-simple.APP"
-"C:\system\apps\qanda-simple\qanda-simple.app"
"..\..\..\epoc32\release\armi\urel\qanda-simple.rsc"
-"C:\system\apps\qanda-simple\qanda-simple.rsc"
```

Figure 3.13 Information needed for generating an installation package

This method completes the implementation of the engine. As expected, the engine contains mostly application-specific code, and it must often be composed from scratch for new applications.

3.5.9 Generating Installation Package

Finally, an installation package must be generated. In the Symbian environment, such packages are referred to as SIS (or SISX in newer systems) packages, and they consist of defining where cross-compiled code resides in the development workstation, and where it is to be installed in the device. In addition, it is possible to include auxiliary files. For instance, in this application, it would make sense to create auxiliary files for questions and answers, and to copy them to a convenient location in the device.

A sample data needed for generating the SIS package for this application is listed in Figure 3.13. After the generation of the package, the size of the application's installation file totals 4 kb. In addition to the minimal contents included in the figure, it is possible to also include additional information, such as installation notes or information on the vendor, in the file. Moreover, in order to support backing up the data of the application, additional information must be provided to register the application's data for backup routines.

3.6 Summary

- The application concept is about connecting new functionality to an already existing platform. The concept defines how applications are integrated with the facilities of the device. This enables the introduction of user-installed applications.

- Application priority is a combination of its execution priority and its importance of being kept in execution.
- Five principal topics form a guideline for application development:

1. well-defined scope of an application,
2. performance,
3. proper UI design,
4. internal representation,
5. communications model.

- Mobile application development often requires an iterative approach for the best UI design and adequate performance.
- Properties of application architectures reflect the assumption on the complexity of assumed applications; the model-view-controller (MVC) pattern is usually applicable as the basis of applications in any environment. It is common that views and controllers are device specific, but that models can be reused in different environments.
- Applications can be packed to different platform-specific installation packages, which have different formats.
- Application infrastructure complexity varies considerably. For example, mobile Java only requires the definition of a single class, whereas Symbian application architecture is based on an elaborated framework on top of which applications are defined. Still, in many cases the parts that the application developer can define are similar, and the differences are only in how and in which places the developer is to inject the actual application code.

3.7   Exercises

1. What would be the main use cases of a multi-user calendar application? What kinds of performance requirements would they imply? Which features form the bottleneck that should be considered when estimating the performance of the application?
2. In what kinds of cases would it be beneficial for applications to share data? How about program code? What kinds of applications should be activated due to environment activity?
3. What would be the minimum application support one must include in a mobile platform? How would that correspond to MIDP Java and Symbian OS applications?
4. Compare application development in MIDP Java and the Symbian OS environment. What kinds of benefits or problems can you find in them?
5. Modify the given sample midlet so that it asks questions in a random rather than in a fixed fashion. Compile it, and install the application to different types of devices (or emulators). Can you find differences in the usability of the application due to the facilities offered by the device?

6. What benefits or problems would occur when implementing a Java environment where both CLDC- and CDC-based profiles could be used? For which kinds of devices would this be an option?

7. What kinds of operations could be offered for managing downloadable applications in mobile devices?

8. In addition to models, which parts of applications could be reused in a different application or device? Why? How would this be visible in designs?

9. Assume a system where applications can be automatically shut down when no execution time is allocated to them within some period. When the application is shut down, it saves its data to the disk, and when the user returns to this application, the saved data is automatically loaded. How does this relate to the use of virtual memory, where an application image is effectively saved to the disk, when no execution time is given to it?

4

Dynamic Linking

4.1 Overview

Dynamic linking, often implemented with dynamically linked libraries (DLL), is a common way to partition applications and subsystems into smaller portions, which can be compiled, tested, reused, managed, deployed, and installed separately. For instance, assuming that a design of an application follows the MVC pattern discussed in Chapter 3, it is sometimes convenient to implement the model as a separate library that can be loaded when necessary. Then, the library can be reused in some other graphical user interface that needs the same functions, for instance, or by applications that integrate the model to a larger context. Furthermore, the use of dynamically linked libraries can also be a way to interact with a given framework, thus allowing a well-defined boundary between two units of software. In particular, extensions and adaptations to an already existing system can often be more elegantly handled using dynamic libraries than with source code, because recompilations of already compiled parts can be avoided and the correct configuration can predominantly be created with binary files.

4.1.1 Motivation

In the sense of program implementation only, it is not usually necessary to divide the program into several libraries or components. However, using dynamically loaded libraries as an implementation technique for allowing several applications to use the same functions has several benefits. In the following, the most important conceptual advantages are listed:

- Several applications can use the library in such a fashion that only one copy of the library is needed, thus saving memory.
- Application-specific tailoring can be handled in a convenient fashion, provided that supporting facilities exist.
- Smaller compilations and deliveries are enabled.

Programming Mobile Devices: An Introduction for Practitioners   Tommi Mikkonen
© 2007 John Wiley & Sons, Ltd

- Composition of systems becomes more flexible, because only a subset of all possible software can be included in a device when creating a device for a certain segment.
- It is easier to focus testing to some interface and features that can be accessed using that interface.
- Library structure eases scoping of system components and enables the creation of an explicit unit for management.
- Work allocation can be done in terms of dynamic libraries, if the implementation is carried out using a technique that does not support convenient mechanisms for modularity.

There can also be other motivations. For instance, a dynamically linked library may be used as a means to manage interface coherence if no other facilities are provided by the implementation infrastructure. Then, applications can be forced to follow a certain interface that has been externally defined, making the interface a separate entity for maintenance and future software evolution. This eases publishing the interface as the underlying implementation can be changed, following the basic principles of encapsulation.

4.1.2 Release Definition Using Dynamically Linked Libraries

Assuming that a system consists of a number of libraries rather than a single application file, a question arises on what constitutes a complete system that must be compiled, installed, and executed. In other words, where does one define what files and auxiliaries are needed for performing compilation and installation for a complete system? We will refer to this task as release definition. In general, release definition is among the first tasks that are to be performed in a development project where several components are needed. Then, the developers know what to aim at, and how the final system is constituted. In contrast, without a release definition, there can be serious ambiguities on what to develop in the first place.

In the simplest form, release definition can consist of a collection of file names in a particular file from which the compilation environment reads them and performs a compilation, like a make file, for instance. However, the more complex the development process and the system to be developed are, the more sophisticated support must be offered. For instance, release definition can be connected to the version control system, and whenever a new version of a component is submitted the system is compiled, resulting in continuous integration of newly completed components.

For obvious reasons, there can be several levels of release definitions. For instance, a project aiming at the development of a mobile device produces a release definition suited for the device, platform developers have a release definition for a particular platform version, and application developers have a release definition for their application. Furthermore, different increments can be defined using release definitions as well, resulting in a number of (internal) release definitions for one project.

A release definition can overlook some parts of the runnable system. Then, missing pieces are assumed to be provided by the platform or by other systems built on top of the release. However, recording this assumption, together with associated version information, can sometimes clarify the development.

4.1.3 Required Implementation Facilities

The software infrastructure needed for implementing dynamically linked libraries is relatively simple. The system provides facilities for dynamic loading and unloading of libraries. Loading the library typically means that it is instantiated in the memory that is accessible by the process that wishes to run the code in the library. However, as program code should not be modified, it is sometimes possible to use shared memory, allowing all processes using the same dynamic library to rely on a common copy, which in some sense provides similar benefits than in-place execution, as superfluous copying can be avoided in both approaches. Obviously, the library should only be unloaded, i.e., removed from the memory, when all the processes using it agree on the removal.

At a more detailed level, instantiating library code to memory in a mobile setting is usually implemented in full, so that both code and variables are instantiated in RAM memory, even if in-place execution, where code is executed directly from ROM, would sometimes be possible as well. Therefore, it is possible to update files by downloading new versions to a location from which dynamically linked libraries will be searched before attempting to load them from ROM. For some particular situations, this can however lead to hard-to-trace errors, if the old and new libraries are not fully compatible in some respect. For instance an application that has worked perfectly with the old version can reveal a number of bugs with the new version due to e.g. some new branches of code that will be executed due to upgrades in the library. In principle, version information could be used as a mechanism to lessen such problems, but in practice many of the problems are unintentional and created even if the developers are expecting that compatibility is not violated.

Obviously, in order to use dynamic libraries in a reasonable fashion, it is advantageous if all code is not needed simultaneously. If that is the case, the benefits of dynamically linked libraries are degraded, because the amount of memory that is needed must be enough for running the whole system in any case. Still, the fact that only one copy of the library is needed, not one per application using it, is an advantage when considering storage space, as already discussed above. In addition, eased development and testing, and fixed partitioning of the system may provide enough reasons for using dynamic libraries. Furthermore, also issues like subcontracting may be considered.

4.1.4 Static versus Dynamic DLLs

While dynamically linked libraries are all dynamic in their nature, there are two different implementation schemes. One is static linking, which most commonly means that the library is instantiated at the starting time of a program, and the loaded library resides in the memory as long as the program that loaded the library into its memory space is being executed. The benefit is that one can use the same static library in several applications. For instance, when relying on MVC as the architecture of an application, one can imagine a model implemented as a dynamically linked library that can be used by several applications.

In contrast to static DLLs, dynamic DLLs, which are often also referred to as plugins, especially if they introduce some special extension, can be loaded and unloaded whenever needed, and the facilities can thus be altered during an execution. The benefit of the approach is that one can introduce new features using such libraries. For instance, in the mobile setting one can consider that sending a message is an operation that is similar to different message types (e.g. SMS, MMS, email), but different implementations are needed for communicating with the network in the correct fashion.

4.1.5 Challenges with Using DLLs

While the abstraction of DLL is commonly used, it has plenty of leaking capabilities. Some of them are related to the way in which dynamic libraries are installed, but some others are related to the way they refer to each other. We will address this topic in the following.

A common risk associated with dynamic libraries is the fragmentation of the total system into small entities that refer to each other seemingly uncontrollably. When more and more dynamic libraries are added to the system, their versioning can result in cases where some applications work with a certain version of a library, whereas others require an update or a previous or newer version. Furthermore, managing all the dependencies between libraries is another source of difficulties, as a collection of compatible components is often needed. In fact, one can create a system where one dynamically linked library loads another that in turn loads further libraries. Managing all these dependencies in application development can be difficult, as the loading time of dynamically linked libraries can seem random due to loading of further libraries.

In addition to dependency management, also other problems result from recursive loading of libraries, since this can result in delays in the execution of the application. The situation is worsened by the fact that in many cases library code is not optimized for the needs of a certain application, but their initialization code can contain parts that are needed by only some of the applications using them. However, for an application programmer it is not in general an option to modify such library code, as it may be delivered in binary format only. Moreover, altering the library for the needs of one application only can be considered impractical, as also other clients of the library should be adapted to the change. This can reveal the implementation techniques of library linking and loading to the user, as well as the amount of loaded libraries.

Another problem is that if a dynamic library is used by only one application, memory consumption increases due to the infrastructure needed for management of the library. In addition, performance overhead can be associated with selecting and loading the right library, which is harmful for obvious reasons. When unloading a static DLL, this seldom is a problem, as the application is often terminated in any case. However, with dynamically loaded DLLs, also this can be considered harmful, since the removal must usually be completed before loading the next library that will then adopt the same role.

In addition to the technical challenges, market requirements lead to the evolution of device types in terms of new features. Often requiring software implementation in mobile devices, such new features can lead to incompatibilities between different versions of the devices, although they in principle are building on the same platform. In order to ensure some measure for compatibility, API releases, where a set of interfaces is promised to remain unaltered for some period of time, can be used for ensuring the continuity of the platform. When changed for another interface, the interface may be declared deprecated for a period of time before its actual removal. This allows time for application programmers to change their implementations that rely on the interface between the two releases.

4.2   Implementation Techniques

Fundamentally, dynamically linked libraries can be considered as components that offer a well-defined interface for other pieces of software to access it (Figure 4.1). As illustrated in the figure, additional information may be provided to ease their use. This is not a necessity, however, but they can be self-contained as well, in which case the parts of the program that use libraries must find the corresponding information from libraries. Usually, this is implemented with standard facilities and an application programmer has few opportunities to optimize the system.

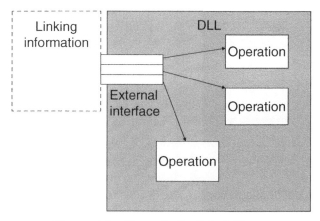

Figure 4.1   Dynamically linked library

Dynamically linked libraries can be implemented in two different fashions. The first way to implement such a system is to create a data structure similar to the virtual function table introduced in Chapter 1, using externally visible operations as the basis for table creation, and use this facility for figuring out the correct memory location to call. In contrast, a more complicated approach can be used, where method signatures are used as basis for linking, where linking is based on language-level facilities, such as class definitions and method signatures for instance.

OffSet-Based Linking. Linking based on offsets is probably the most common way to load dynamic libraries. The core of the approach is to add a table of function pointers to the library file, which identifies where the different methods or procedures exported from the dynamically linked library are located, thus resembling the virtual function table used in inheritance. Then, when the library is used, the compilation can be performed against the exported functions using offsets as identifiers in generated code. The benefits of the approach are obvious. As calling of exported functions is performed directly through a function pointer, the result is suitable with respect to performance. However, there are several shortcomings. Firstly, the compiler can only perform restricted optimization, as for instance inlining can cause problems for linking; how to create an entry point to the inlined function? Similarly, additions to the library can invalidate some of the applications using it, because the table used for linking when using the library and the table inside the library may become different due to failed compilations, or forgotten updates or registrations. The situation is illustrated in Figure 4.2.

Signature-Based Linking. In contrast to offset-based linking of dynamically linked libraries, also language-level constructs, such as class names and method signatures,

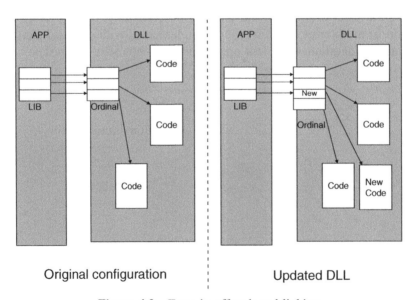

Figure 4.2   Error in offset-based linking

can be used as the basis for linking. Then, the linking is based on loading the whole library to the memory and then performing the linking against the actual signatures of the functions, which must then be present in one form or another. This results in a more fail-safe but more elaborate design, where dynamic linking may become a bottleneck when starting (or running) applications if a long chain of libraries is loaded. Moreover, memory consumption is increased considerably, because all libraries must include additional information regarding the signatures of methods they offer to their clients. Similarly, all method calls made by some other code using the services of the dynamic library must include usable information regarding the signature as well, further contributing to the increased memory footprint.

Based on the above, in addition to the way in which dynamically linked libraries are used, also their implementation facilities are capable of leaking. However, depending on the implementation scheme, the users of the libraries will experience leaking in a different fashion.

4.3   Implementing Plugins

Plugins, which dynamically loaded dynamically linked libraries are often referred to as, especially if they play a role of an extension or specialization, are a special type of DLL that enable differentiation of operations for different purposes at run-time. They usually implement a common interface used by an application, but their operations can still differ at the level of implementation. As already mentioned, one could implement different plugins for different types of messages that can be sent from a mobile device, for example.

4.3.1   Plugin Principles

Plugins take advantage of the binary compatibility of the interface provided by a dynamically linked library, as illustrated in Figure 4.3. The important concepts of a plugin are the interfaces they implement, and the implementations they provide for interfaces. The interface part is used by applications using the plugin for finding the right plugins, and the implementation defines the actual operations. Commonly some special information regarding the interface is provided, based on which the right plugin library can be selected.

When a plugin is selected for use, its implementation part is instantiated in the memory similarly to normal dynamically linked libraries. Obviously, it is possible to load and unload several plugins for the same interface during the execution of an application, depending on required operations.

While no special support apart from common DLL loading and unloading is an absolute necessity for plugins, software infrastructure can offer special support for them, which eases the use of the feature. For instance, there can be operations for identifying interfaces based on interface identifier. Similarly, there can be an

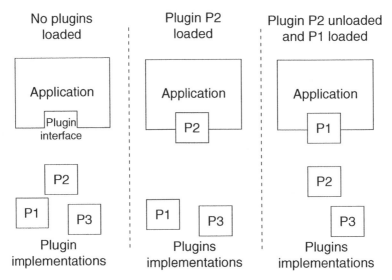

Figure 4.3 Using plugins

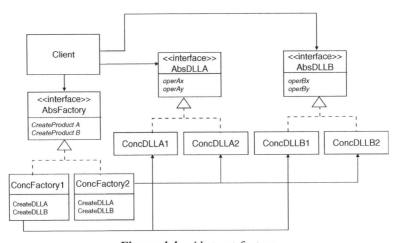

Figure 4.4 Abstract factory

identifier for implementations as well, potentially helping in the selection of the right implementation for some interface.

One applicable solution for the implementation of plugins is the abstract factory design pattern introduced by Gamma et al. (1995). The pattern is illustrated in Figure 4.4, in which prefixes Abs and Conc refer to abstract and concrete elements of the design, respectively. In the pattern, an interface is defined that is accessible to an application. However, no implementation is provided for the interface, but the application must instantiate one using a special factory operation similarly to Symbian OS's `NewL` and `NewLC` methods discussed earlier. The operation allows parameters that guide in the selection of the convenient implementation, thus offering more options for the application to select a well-suited plugin. Several types of approaches are available, where the programmer has different responsibilities. In some cases, the programmer is responsible for all the operations, in which case all the plugins are just plain dynamically linked libraries from which the programmer selects one. In other cases, sophisticated support for plugins is implemented, where the infrastructure handles plugin selection based on some heuristics, for instance. However, the applicability of this approach is restricted, as finding general-purpose heuristics is hard, if not impossible.

The idea of plugins can be applied in a recursive fashion. This means that a plugin used for specializing some functions of the system can use other plugins in its implementation to allow improved flexibility. For instance, a messaging plugin discussed above can use communications plugins to transmit the message over GPRS or Bluetooth in a fashion that is invisible to the application using the facility. One should however note that while the earlier problems on creating a recursive sequence of dynamically linked libraries can be difficult to manage when using regular DLLs, similar use of plugins can be considered even more complex due to the more dynamic relation between the libraries. Therefore, even more care should be given to ensure the creation of a manageable configuration.

4.3.2 Implementation-Level Concerns

While plugins in principle are a simple technique, there are several implementation details that must be considered before applying the approach. In the following, we elaborate some practicalities of using them in applications.

To begin with, a framework is commonly used for loading and unloading plugins. This implies a mechanism for extending (or degenerating) an application on the fly when some new services are needed.

Secondly, in order to enable loading, facilities must be offered for searching all the available implementations of a certain interface. This selection process is commonly referred to as resolution. In many implementations, a default implementation is provided if no other libraries are present, especially when a system plugin that can be overridden is used.

Finally, a policy for registering components and removing them is needed in order to enable dynamic introduction of features. This can be based on the use of registry files, or simply on copying certain files to certain locations, for instance.

One should also note that using the library that has already been loaded is not always intuitive. As an example, let us again consider the case where different message types and associated messaging features have been implemented using plugins. When composing a message, the user is then prompted to determine the type of the message she is planning to send, which is, say MMS in this example. The device loads the right plugin, and allows the user to compose the message, which is then sent using the plugin. When the message is received, the receiving device must recognize that the incoming message requires an MMS plugin, which is then loaded, and used for displaying the message. The confusing part is what happens if the user decides to reply to the MMS. As the MMS plugin is already loaded, the user may not be prompted about the message type but the present plugin is used automatically, which is the fastest and probably the most convenient way. As a result, it may be impossible to respond to an MMS using an SMS even if it would make more sense to send a textual response only.

4.4   Managing Memory Consumption Related to Dynamically Linked Libraries

As already discussed in Chapter 2, memory consumption forms a major concern in the design of software for mobile devices. At the same time, a dynamically linked library is often the smallest unit of software that can be realistically managed when developing software for mobile devices. Therefore, in this section we introduce some patterns introduced by Noble and Weir (2001) for managing memory consumption at DLL level. One particular detail that should be considered is that when managing memory consumption, some of the available memory will necessarily be allocated for implementing management routines.

4.4.1   Memory Limit

Setting explicit limits regarding memory usage for all parts of the system is one way to manifest the importance of controlling memory usage. Therefore, *make all dynamically linked libraries (and other development-time entities) as well as their developers responsible for the memory they allocate.* This can be achieved, for instance, by monitoring all memory reservations made by a library or a program. This can be achieved, for example, using the following routine, where MyLimit is the maximum amount of memory the library (or subsystem) can allocate and myMemory refers to the amount of allocated memory.

```
void * myMalloc(size_t size)
{
#ifdef MEMORY_LIMITATION_ACTIVE
    if (myMemory + size > myLimit) return 0;
    else { // Limit not reached.
        void * tmp = malloc(size);
        // Update myMemory if allocation succeeded.
        if (tmp) myMemory = myMemory + size;
```

```
        return tmp;
    }
#else
    return malloc(size);
#endif
}
```

While the above procedure only monitors memory space used for variables, the same approach is applicable to programs' memory consumption as well. Then, the role of the approach is to place developers' focus on memory consumption during the design. Furthermore, designing memory usage in advance creates an option to invest memory to some parts of a system for increased flexibility, and to optimize for small memory footprint on parts that are not to be altered.

In order to be able to give realistic estimates for future releases when setting memory limits for them, one should maintain a database of memory consumption of previous releases to monitor the evolution of the consumption. Moreover, more precise estimates of the final results can be obtained by also including estimates made at different phases of the development project into the database, which can be used for evaluating the effect of errors in estimates made in the planning and design phases.

4.4.2   Interface Design Principles

As with many designs, there is no single fundamental principle that would always overrule in the design of interfaces. Instead, one can advocate a number of rules of thumb that are to be considered. In the following, we will address such issues.

Select the right operation granularity. In many cases, it is possible to reveal very primitive operations, out of which the clients of a dynamically linked library can then compose more complex operations. In contrast, one can also provide relatively few operations, each of which is responsible for a more complex set of executions. A common rule of thumb is to select the granularity of the visible interface operations so that they are logical operations that a client can ask the library to perform, and not to allow setting and getting of individual values (overly simplistic operations) or simply commanding the library to doIt() (overly abstract operations), for instance.

Allow client to control transmissions. This allows implementations where clients optimize their memory and general resource situation in their behaviors, whereas if the service provider is in control, all the clients get the same treatment, leaving no room for special situations on the client side.

Minimize the amount of data to be transmitted. For individual functions that call each other this means that the size of the stack remains smaller. Furthermore, there will be less need for copying the objects that are passed from one procedure to another. For cases where process interface restricts communication, minimizing the amount of data will lead to a design where less data is passed between processes.

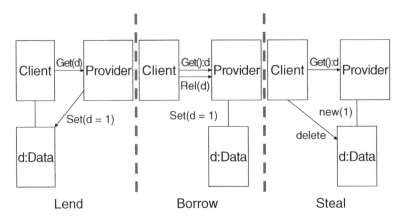

Figure 4.5 Ways to transmit data from a service provider to a client

This in turn will mean less message passing. Even if processes communicate via shared memory, minimizing the amount of data that is passed usually leads to improved performance and helps to control memory consumption. Finally, when data is transmitted over the network, minimizing data transmissions usually result in increased performance.

Select the best way to transmit the data. There are three fundamentally different ways to transmit data, referred to as lending, borrowing, and stealing (Figure 4.5). They are described in the following in detail.

1. *Lending.* When a client needs a service, it provides some resources (e.g. memory) for the service provider to use. The responsibility for the resource remains in the client's hands.
2. *Borrowing.* When a client needs a service, it expects the service provider to borrow some of its own resources for the client's purposes. The client assumes the responsibility for the deallocation, but uses the service provider's operation for this.
3. *Stealing.* The service provider allocates some resources whose control is transferred to the client. The owner of the resource is changed, and the client assumes full ownership, including the responsibility for the deallocation.

When making the decision, there are several viewpoints to consider. For instance, it may be relevant whether it is the client or the service provider that takes the initiative. For instance, in some cases the service provider may already have an existing copy of the data item that only one client is accessing, and it may be easiest to simply pass the copy to the client and forget about it, assuming that no other client needs the data. Similarly, when a number of clients will be accessing the same data, different decisions may be taken.

4.4.3 Merging Elements

In general, merging elements as a trick to save memory is obviously deprecated. However, in cases where different relevant design choices are available, as well as in cases where extreme measures are needed, one may need to consider even such actions. However, they probably should be used as the last resort only, when no other action saves the program, as usually it is easier to find better data structures by other means.

Merging packages and dynamically linked libraries. By merging packages and dynamically linked libraries, some packaging infrastructure can be saved. Then, when the contents of the packages refer to each other, no package specification is needed. The downside is that loading will be performed on a coarser scale. The same applies to necessary management functions.

Flattening structural hierarchies. Hierarchies associated with generated structures, such as packages and inheritance hierarchies, consume memory that can be saved to some extent. With packages of MIDP Java, for instance, it is possible to reduce the number of identifiers that are needed for dynamic linking. With inheritance, flattening hierarchies directly reduces the number of virtual function tables needed for representing the run-time configuration. Moreover, functions that are not over-ridden will be visible in several virtual function tables, which further encourages reducing their number.

Embedding objects. By embedding small objects (or individual data fields such as pointers) into bigger objects, less overhead is associated with objects, as an object tag is always needed per object, as illustrated in Figure 4.6. Additional, and often more important, benefits can be gained due to the simplified allocation strategy. For instance, assuming that a list is implemented in a fashion where all the nodes are objects that bear a reference to the next node and to the actual data item, a revised implementation where data is included in the node results in improved memory

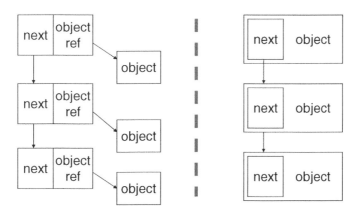

Figure 4.6 Embedding pointers

allocation. Furthermore, as the node then becomes a bigger entity, there are also benefits regarding cache usage.

4.5   Rules of Thumb for Using Dynamically Loaded Libraries

In principle, any unit of software can be implemented as a separate library. In practice, however, creating a complex collection of libraries that depend on each other in an ad-hoc fashion is an error that is to be avoided at all cost. In the following, we list some candidate justifications for creating a separate dynamically linked library out of a program component.

- *Reusable or shareable components* should be implemented using dynamically loaded libraries, as otherwise all the applications that use the components must contain them separately. This in turn consumes memory. In addition, sharing can take place in a form where the same model, implemented as a dynamically loaded library, is reused in several different devices that require a specialized user interface for each device.
- *Variation or management point* can be preferable to implement in terms of dynamic libraries. This makes variation or management more controlled, as it can be directly associated with a software component. Moreover, the library can be easily changed, if updates or modifications are needed. For instance, interfaces can be treated in this fashion even if the underlying infrastructure (e.g. C++) would not offer such an option. Management can also be associated with scoping in general, allowing that a certain module can be evolved separately.
- *Software development processes* such as automated testing may require that all the input is in a form that can be directly processed. Then, performing the tests may require that binary components are delivered.
- *Organizational unit* can be authorized to compose a single library that is responsible for a certain set of functions. Requesting a library then results in a separate deliverable that can be directly integrated into the final system.

The above list is by no means exhaustive, and new rationale for using dynamically loaded libraries can be found when considering case-dependent rationale. However, similar justification should be provided for any decision to use a separate library, and such should never be made due to a single decision by a programmer.

4.6   Mobile Java and Dynamic Linking

Infrastructure associated with dynamic linking in Java is simple: all classes are dynamically linked. Loading and linking them is a built-in feature of the environment, which is necessarily based on signatures.

Table 4.1   Average class file content in CLDC library classes

Item	Percentage
Metadata	45.4%
Strings	34.8%
Bytecodes	19.1%
Class field data	0.4%
Instance field data	0.4%

Table 4.2   Effect of compression and different application structures

Format	14 Classes	1 Class
No compression	14 019	7467
No compression, obfuscated	13 093	6929
JAR	10 111	3552
JAR obfuscated	10 048	3540
Pack	4 563	3849
Pack.gz	2 568	2235

When using MIDP Java, all necessary classes are loaded when an application is initiated. At the same time, some measures are taken to check that the application will not run into invalid references, for instance, which is assisted by the build mechanism described in Chapter 8. Because linking is performed using the actual names, the profile of memory consumption also includes a lots of metadata (Table 4.1[1]) (Hartikainen, 2005).

A related issue is that in many cases, classes contain the same information over and over again. The reason is that for instance method names are often the same (or similar) in different classes, in particular those that have the standard denotation in bytecode, such as constructors, for example. Therefore, using different formats can considerably alter memory consumption. As an example, Table 4.2 introduces sizes of an application implemented as 14 and as 1 class, and the effect of different compressions in accordance to Hartikainen et al. (2006). As an additional example of the effect of compression, Table 4.3 introduces the effect of different compression based on measurements made by Hartikainen et al. (2006) using CLDC 1.1 library classes as the data to be compressed. As examples, we use JAR, Pack200, JXE, gzip, and tar. Being able to use a more efficient compression when downloading applications would result in faster operations. Unfortunately, this requires modifications to the standard.

---

[1] Metadata includes up to 50% debugging data, whose contribution should also be taken into account.

Table 4.3 Effect of compression to library classes

Format	Size (bytes)
Classes	111 694
JAR	76 517
Pack	43 233
Pack.gz	23 791
tar.gz	46 099
JXE	104 919

For faster startup of applications, many implementations contain some prelinking of standard libraries that will be used by most, if not all, applications in the virtual machine. Directly embedding the libraries in the core virtual machine in a precompiled and prelinked form leads to larger memory consumption, because Java bytecode is relatively efficient and all the libraries are not always needed. However, application startup time can be enhanced considerably, as no actual loading needs to be performed.

Finally, when using MIDP Java, only the default class loader can be used, which restricts the possible options of a programmer, but reduces memory footprint of the virtual machine, allows a simpler implementation, and simplifies the underlying security scheme. Another simplification is that no opportunity to implement plugins is offered.

4.7 Symbian OS Dynamic Libraries

Symbian OS dynamic libraries are a realization of the offset-based mechanism described earlier. When libraries are used, they are sought from the same disk as the process with which the libraries will be linked has been started. Therefore, in-place execution can be used, if the hosting process resides in ROM.

4.7.1 Standard Structure of Dynamically Linked Libraries

The general structure of dynamically linked libraries in the Symbian environment is based on static offset-based linking, and the export interface table is referred to as ordinal. There are several key concepts that are related to the topic, including the library structure, applied implementation techniques, and binary compatibility. We will address these topics in more detail in the following.

Structure of Dynamically Linked Libraries

Because the main implementation language of Symbian OS programs, C++, as such does not support the separation of a DLL interface, special macros are used for defining the interface when composing Symbian programs. First, macro IMPORT_C is to be used in header files to determine the methods that are to be made visible from the library. Then, macro EXPORT_C is to be used in the actual file when the method is a virtual method.[2] Of particular importance is the fact that the above macros are only needed when composing code for the actual device. The emulator of the Symbian environment does not need the macros. Additional consideration must be given to the creation of all files associated with a DLL interface, freezing. Once performed, auxiliary files will be generated that enable linking of the library to other components. For obvious reasons, the basis of forming the auxiliary file must be based on the same (or compatible in terms of interface generation) code as will be used in the application to access the operations.

Implementing a Sample Model as a Dynamically Linked Library

Let us consider next that we wish to implement the engine of the sample application given in Chapter 3 as a separate dynamically linked library. The header and the code of the library are given in files qaeng.h and qaeng.cpp, respectively.

The header of the dynamic library is almost the same as the one given earlier. The only difference is that a number of IMPORT_C macros are needed, resulting in the following listing:

```
#include <e32std.h>
#include <e32base.h>
#include <badesca.h> // Array use from badesca.h.

class CQAEng : public CBase
    {
public:
    IMPORT_C static CQAEng* NewL();
    IMPORT_C ~CQAEng();
    IMPORT_C void Select();
    IMPORT_C TPtrC Answer();
    IMPORT_C TPtrC Question();
    IMPORT_C TBool Used();
    IMPORT_C void Reset();
protected:
    void ConstructL();
    CQAEng();
private:
    TInt iAnswer, iQuestion;
    TBool iUsed;
    CDesCArrayFlat * iQuestions, * iAnswers;
};
```

---

[2] The minimal definition for exported functions differs slightly in different sources and contexts. Here, we give the most general definition.

Finally, in the implementation, we again need a number of macros, this time EXPORT_C. This leads to the following definition for NewL, for example:

```
EXPORT_C CQAEng * CQAEng::NewL()
    {
    CQAEng* self = new (ELeave) CQAEng;
    CleanupStack::PushL(self);
    self->ConstructL();
    CleanupStack::Pop();
    return self;
    }
```

Similar treatment is given to other methods that will be visible from this library, covering essentially those that had macro IMPORT_C above.

In addition, we need to introduce the necessary entry point (E32Dll) to the dynamically linked library, which is not unlike the one that was introduced already in connection with the concept of applications:

```
EXPORT_C TInt E32Dll( TDllReason )
    {
    return KErrNone;
    }
```

Method ConstructL and constructor remain unaltered, as they are visible at the DLL interface through associated factory functions only.

4.7.2 Managing Binary Compatibility

In order to implement applications that run on several different devices, preserving binary compatibility of libraries is an important issue to consider. Fundamentally, the main principle is to ensure that new methods are only added to the end of the ordinate, never in between the methods that have been used in older versions of the libraries. For obvious reasons, performing this is only a concern if a dynamic library is used by several applications, and the applications cannot be upgraded at the same time as the library. In particular, this therefore is a concern of platform manufacturers, as all the platforms should ideally remain backward compatible.

Achieving such a strategy in an implementation is not straightforward in practice, however, but requires consideration of the effects of changes at the level of generated code. A number of more elaborate binary compatibility rules have been introduced by Stitchbury (2004):

- *Do not change the size of a class object*. Exceptions to this rule are cases when one can ensure that the class is not externally derivable, all allocations of objects of this class take place in one's own component, and the class has a virtual destructor.

- *Do not remove anything accessible.* Otherwise clients of the library can access wrong variables.
- *Do not rearrange accessible class member data.* This modifies their representation in memory, and thus lets client programs access wrong variables.
- *Do not rearrange the ordinal of exported functions.* This will lead to calling wrong methods because ordinal is the basis for linking.
- *Do not add, remove, or modify the virtual functions of externally derivable classes.* This will break binary compatibility for obvious reasons.
- *Do not re-order virtual functions.* C++ implementations commonly use the order in which virtual member functions are specified in the class definition as the only factor that affects now they appear in the virtual function table.
- *Do not override a virtual function that was previously inherited.* Performing this alters the virtual function table of the class. Then, clients compiled against the old and new virtual function table behave differently.
- *Do not modify the documented semantics of an API.*
- *Do not remove* `const`.
- *Do not change from pass by value to pass by reference, or vice versa.* Generated code will be totally different, resulting in errors in execution.

In addition to the above cases, which clearly break the binary compatibility, Stitchbury (2004) also defines some principles for developing flexible dynamic libraries. While they may not be a risk as such, common needs of evolution can lead to difficulties, especially when considering the long-term future of the system.

- *Do not inline functions.*
- *Do not expose any public or protected member data.* This in general is considered a bad practice.
- *Allow object initialization to leave.*
- *Override virtual functions that are expected to be overridden in the future.* This will not lead one to trust on the underlying behavior.
- *Provide 'spare' member data and virtual functions.* As adding of methods and data can be difficult once the interface and the contents of an object are fixed due to compatibility reasons, precautions can be taken to ease maintenance.

Obeying these simple guidelines leaves the developer the option to perform the following modifications (Stitchbury, 2004):

- *API extensions can be introduced.* New operations can be added to interfaces, as long as they do not change those that already exist.
- *Private internals of a class can be modified.* They are local to the class, so it can implement them in any desired fashion.
- *Access specification can be relaxed.* Old clients that can survive with a more restricted access will not be harmed with this.

- *Pointers can be substituted with references and vice versa.*
- *The names of exported non-virtual functions can be modified.* While this does not cause technical problems because the ordinal is not altered, practical consequences need also be considered, in particular semantics.
- *Input can be widened and output can be narrowed.* Again, old clients are not affected with this change.
- *Specifier* `const` *can be applied.*

4.7.3 ECOM Component Model

ECOM component framework is a Symbian OS implementation for plugin components. Used in devices from version 8.0 onwards, it provides facilities for using polymorphic dynamically linked libraries. Figure 4.7 illustrates clients, interfaces, implementations for the interfaces, and the role of the ECOM framework, assuming that some kind of crypting software (`CCryptoSW`) is to be implemented as a plugin (Stitchbury, 2004).

An ECOM interface (`CCryptoIF` in the figure) has the following characteristics (Stitchbury, 2004):

1. It is an abstract class with a set of one or more pure virtual functions.
2. It must provide one or more factory functions to allow clients to instantiate an interface implementation object.
3. It must provide means for its clients to release it (for example a destructor, `Release`, `Close`, or `Deallocate`).
4. It has a `TUid` data member, which can be used internally to identify an implementation instance for cleanup purposes.

In order to use the framework, one must define factory methods that are used for instantiating plugins. This can be performed using either the UID of the

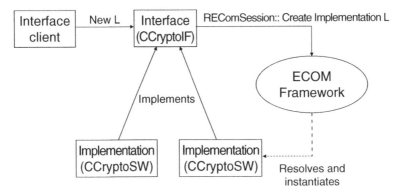

Figure 4.7 ECOM overview

implementation as a reference, or using a customizable ECOM resolver to find a default implementation (Stitchbury, 2004).

Actual components are implemented using the same mechanism as all dynamically linked libraries. However, an additional instantiation function is needed that the library registers to the ECOM framework. This can (and in many cases should) be a `NewL`, for instance. A plugin registers this method by exporting a standard function `ImplementationProxyGroup`. which returns a pointer to an array of instances of class `ImplementationProxy` objects. This is the only method that the library exports. Each `ImplementationProxy` object represents a single implementation class, and contains the `TUid`, which identifies the implementation and a pointer to its instantiation method (Stitchbury, 2004).

Polymorphic dynamically linked libraries can also be implemented without using ECOM. However, then the whole responsibility must be assumed by the developer in full. Such an implementation was commonly used before the introduction of ECOM facilities. Obviously, for future use, ECOM is the preferred mechanism in most applications and facilities.

4.8   Summary

- Dynamically loaded libraries enable memory savings, because only those libraries that are needed must be present. However, as the implementation of the necessary library infrastructure requires memory, and loading of dynamic libraries takes time, their use must be carefully considered.
- Release definition is the process that defines which files are included in a completed system.
- Linking can be based on method signatures, like in Java, or on more static approaches, where for instance offset of a function is used, like in Symbian.

  - Using method signatures is more flexible but there is considerable memory overhead as signatures must be available even after the compilation.
  - Using offsets offers better performance but can lead to complex error cases if an offset mismatch occurs.

  This may also have an impact on the size of units; in Java all classes are linked dynamically, whereas in Symbian OS dynamically linked libraries are explicitly composed in terms of classes and associated export interfaces.
- Plugin implementations are special types of dynamically loaded libraries whose interface is the same but whose implementations may vary. This enables different types of adaptation in a controlled and sophisticated fashion.
- Plugin implementation requires

  1. a common interface definition, which provides an abstraction for the operation,
  2. an implementation for the interface; in many cases several implementations are allowed for versatility,

3. a framework for selecting, loading, and unloading the different implementations.

- Dynamic libraries can be used as a technique for specializing a framework in a controlled fashion.

4.9 Exercises

1. Assume that the memory limit pattern is followed in the design of a system consisting of dynamically linked libraries and programs that load and unload DLLs when executed. Should the sum of planned memory consumption be more, equal, or less than the actual amount of memory in a device? Why? What are the consequences of the different choices?
2. What kinds of problems are possible when a dynamically linked library and an application are separated in the device? What kinds of precautions could be made? How do signature- and ordinal-based approaches differ in this respect?
3. For what purposes can one assume plugins to be used in common applications of mobile devices?
4. What if one needs several plugins at the same time? For instance, how would an implementation be composed, where upon receiving a MMS, one wishes to reply with an SMS? What kinds of problems does such a design contain?
5. What kinds of methods would one need when defining an interface for a generic messaging (SMS, MMS, email) plugin? Would one class whose instantiation is required be enough, or should a group of components be constructed when using the factory method?
6. What kinds of errors can occur if the interface of a dynamically linked library relying on the use of an ordinal is incorrectly frozen?

5

Concurrency

5.1 Motivation

In Chapter 3, applications running inside mobile devices were considered reactive. This, however, can be generalized to hold at the device level as well, since it is the device that is assumed to provide responses, not the applications only. Moreover, it is not solely applications that receive and respond to events, but it is a joint effort of several components of the device. These components can include hardware, low-level software directly associated with it, or some other piece of software.

As the result of the above, the software run inside a mobile device can in general be taken as an event handler, which simply responds to the events received from the outside world. Because the outside world behaves nondeterministically from the viewpoint of the device, the device must be enabled to react to several incoming requests at the same time. This calls for an implementation that respects this requirement, and allows concurrent response to several simultaneous incoming events.

The implementation mechanisms for concurrent programming are more or less standardized. As already discussed, threads and processes give boundaries for managing executions and resources. In addition, some mechanisms are available to ensure that several threads do not modify the same variables at the same time. In the following, we will address such issues in more detail.

5.2 Infrastructure for Concurrent Programming

When programming a system where some input is generated by the environment and requires immediate reaction whereas other input leads to extensive executions, parallel processing is usually needed. Then, based on the priority of the thread, the most important stimuli will be handled first. Fundamentally, three different cases can be considered:

1. Executions are unknown to each other. However, they can still affect each other by competing for the same resource, like processor execution time.

2. Executions are aware of each other indirectly. For instance, they may have some common shared resource via which they cooperate.
3. Executions communicate with each other directly.

Such executions are commonly implemented using threading, which we will address in the following.

5.2.1 Threading

Threads that wait for stimuli and react to them are a commonly used mechanism for creating highly responsive applications. This allows incoming stimuli to initiate operations, which in turn can generate new stimuli for other threads, or perhaps more generally lead to the execution of some procedures by the threads themselves. The cost of using threads is that each thread essentially requires memory for one execution stack, and causes small overhead during scheduling.

As already discussed in Chapter 1, threads can be executed in a pre-emptive or non-pre-emptive fashion. The former means that the thread that is being executed can be interrupted, and another thread can be selected for execution. The latter implies that once a thread is being executed, its execution will only be interrupted for another thread when the executing thread is ready. Low-end mobile devices, optimized for low costs in their design, can be implemented using a non-pre-emptive scheduling policy, where all executions take place following a pre-planned order of execution that satisfies all the deadlines. In more sophisticated mobile devices, pre-emptive executions based on thread priorities are used. Such an approach offers simplified facilities for the application programmer, as the threads need not consider their executions. Instead, the system selects the thread to be executed next according to different algorithms. Usually, priorities are introduced that determine the order of execution, i.e., threads whose priorities are higher are selected for execution before considering lower-priority threads.

5.2.2 Inter-Thread Communication

While threads are a mechanism for creating executions, in order to accomplish operations at a higher level of abstraction, several threads are often needed that cooperate for completing a task. For example, establishing and maintaining a phone call requires the cooperation of a user interface, communication protocols, radio interface, a microphone, a speaker, and a unit that coordinates the collaboration. This cooperation requires inter-thread communication.

There are several mechanisms for making threads communicate. In the following, we address the most common approaches that can be used in mobile devices.

Shared Memory

Probably the simplest form of communication is the case where threads use a shared variable for their communication. In most cases, the access to the variable must be implemented such that threads can temporarily block each other, so that only one thread at a time performs some operations on the variable. In general, such operations commonly address memory that is shared by a number of threads, and blocking is needed to ensure that only one thread at a time enters the critical region, i.e., a part of code where only one (or some predefined number of) thread should enter at a time and in which atomic changes to a number of variables can be made. The situation can be simplified by using a convention where only one-way communication is used, i.e., only one thread is able to set the value of a certain variable, and other threads can only read the value. Then, no locking is needed for accessing the variable, assuming that there is no other reason for atomicity.

For more complex cases, semaphores are a common technique for ensuring that only one thread at a time enters the critical region. In the classical definition, semaphores offer two operations, P and V. These operations decrement and increment the value of a semaphore counter, whose initial value determines how many threads are allowed to pass the P operation. Further threads will stop and wait in P until the next V is executed by some other thread.

There are also more abstract ways to implement a function similar to the above. Signal-wait operations are commonly used for inter-thread communication, and they make it possible for a thread to wait for a certain signal before proceeding beyond a certain point in its execution. Furthermore, monitors are a programming language level abstraction for implementing cooperation of threads, where access restrictions can be given in operations.

Message Passing

Message passing is another commonly used mechanism for implementing cooperation between threads. In this type of an approach, the idea is that threads can send messages to each other, and that kernel will deliver the messages. In other words, the architecture of such a system resembles message-dispatching architecture, where the kernel acts as the bus, and individual processes are the stations attached to it.

In many cases, message passing is in principle hidden from the developer. Then, the developer can simply call a simple API routine, and the infrastructure handles the actual execution. However, side-effects of the message passing system often become visible to the developer when certain types of interactions take place in parallel. For instance, it may be so that the execution of a thread is blocked until the thread receiving the message responds (synchronous communication), or that in order to receive the reply a special operation must be called, which register the thread to receive the result (asynchronous communication). We will return to such topics in the following chapter.

5.2.3 Common Problems

Concurrent programs can fail in three different ways. One category of errors in mutual exclusion can lead to locking, where all threads wait for each other to release resources that they can reserve. Secondly, starvation occurs when threads reserve resources such that some particular thread can never reserve all the resources it needs at the same time. Thirdly, errors with regard to atomic operations, i.e., operations that are to be executed in full without any intervention from other threads, can lead to a failure. The two first problems can be solved by introducing a watchdog mechanism that ensures that all threads are executed by adding a simple call to a monitoring routine to all threads. However, with respect to atomicity, it is harder to implement a safety mechanism, although additional sanity check computations and data can be used.

From the perspective of a programmer, the introduction of concurrent executions inevitably leads to further complexity. Firstly, design is more difficult, because several parallel paths must be considered. As a result, the order of operations that is observed to take place can vary from one execution to another. As a consequence, also debugging becomes complicated, because errors are not repeatable; instead, executions can take place in different order at different times, of which only one path may lead to a failure. As in many cases the most rational starting point is to be able to reproduce an error; these non-repeatable errors can be virtually untraceable.

Another design-level problem is about design practices. As mutual exclusion can result in delays (one thread must wait until some other thread has exited the critical section), it is common that performance of the first version is improved by reconsidering mutual exclusion, which often indeed is overly safe in the first version. However, tampering with such detail without thoroughly understanding the logic of the system can seriously violate the design.

As concurrency can be considered as a conceptually difficult topic, and, furthermore, mobile devices often contain only one application processor that is shared by all applications, one way to deal with problems related to mutual exclusion is to fake concurrency in a fashion where the design can be eased.

5.3 Faking Concurrency

In addition to the above difficulties, threads are a relatively heavy implementation mechanism in some cases, especially when several operations can occur at the same time. Then, each operation would basically require a thread that waits at a semaphore or reception operation of a message, resulting in excessive threading in applications. In addition, the use of threads can be considered error-prone, because repeating errors that occur due to for example mismatches at critical regions is next to impossible.

In order to avoid the above problems, it is often enough to enable pseudo-parallel executions where for example only one thread is responsible for all the operations, even if the programmer is allowed to write code as if there are several parallel executions taking place. This effectively leads to serialization of executions. Such serialization can be implemented in a fashion that offers several positive consequences:

- Non-pre-emptive processing reduces overhead. Executions are not interrupted by other operations, resulting in better performance in the sense of completed executions, because the overhead of stopped and resumed executions is eliminated. Instead of defining priorities of execution per thread in a pre-emptive fashion, the event handler determines the priorities in a non-pre-emptive fashion where each event is completely handled before the next event handling is initiated. Effectively, this means that event handling is serialized.

- No need to consider critical regions, as the programmer can design things so that no such issues arise. Moreover, only one event at a time will be under handling, which simplifies this.
- Simplified concurrency scheme leaves less potential for design and programming errors. For instance, mutual exclusion can be made obsolete if only one thread executes the operations associated with the events.
- Simplified implementation in terms of necessary facilities can lead to savings in development effort.
- Porting can be eased, assuming that the same application is to be used in different devices with different operating systems that offer different facilities for a programmer. Then, it is enough to implement one thread and its facilities instead of porting all the necessary infrastructure of concurrent programming.
- It is easier to consider the net effect of the total system, because for instance only one operating system thread can be allocated to serve all the events. Then, even if a lot of computations are introduced, high-priority threads of the operating system are not affected and it is still easy to estimate required operating system resources. This effectively helps in preventing overloading.

However, taking the gains from the above list does not come for free, but there is a penalty hidden in the use of serialized executions as well. If an operation cannot be interrupted for another, a more urgent operation may have to wait for the completion of a less important one. Furthermore, optimizing the serialization in one thread can lead to degenerated performance for the application. Additional consideration must be given to long-lasting executions, which may require the implementation of some kind of a continuation mechanism.

Finally, an important choice to be made when composing a design is to decide whether there will be support for faked concurrency in the implementation infrastructure, or whether the programmers will be responsible for it. In the first case, one can write applications as if concurrent executions would be available using primitives described above, and the infrastructure will be responsible for optimization. Then, the programmers can use data types and operations commonly associated with threads, but the run-time environment will be responsible for performing the serialization. If the programmer takes the responsibility for serialized parallelism, the design will inevitably reflect the selected implementation mechanism, thus adding complexity to the design of the resulting system, at least potentially. In practice, an implementation for such design can be composed by allowing asynchronous request of operations, where the execution environment generates an event upon completion.

This event can then be handled with a mechanism not unlike that introduced for event handling in Chapter 3.

For obvious reasons, real-time critical operations can hardly benefit from faking concurrency. Therefore, some systems allow both serialized form of pseudo-parallel executions as well as full threading. Still, even in this case, native threads will be scheduled for the execution by the underlying operating system, which in fact is only a more low-level mechanism for serializing the execution when only one application processor is used.

5.4 MIDP Java and Concurrency

While Java again in principle hides the implementation of the threading model from the user, its details can become visible to the programmer in some cases. In particular, the developers must consider the following issues.

5.4.1 Threading in Virtual Machine

In the implementation of the Java virtual machine, one important topic is how to implement threads. There are two alternative implementations, one where threads of the underlying operating system are used for implementing Java threads, and the other where the used virtual machine is run in one operating system thread, and it is responsible for scheduling Java threads for execution. The latter types of threads are often referred to as green threads; one cannot see them from green grass. In many ways, the scheme can be considered as a sophisticated form of event-based programming, because in essence, the virtual machine simply acts as an event handler and schedules the different threads into execution in accordance to incoming events.

In general, threads that are directly associated with the operating system's threads can be considered better for performance, because the threads get execution time from the scheduler of the operating system. However, porting of the system to different operating systems becomes more complex when threads are intimately connected to the operating system particularities. An additional source of complexity is that in addition to porting, threads must be manageable by the virtual machine when exceptions take place and some cleanup must be performed.

As a result of the above complexities, green threads form an attractive implementation for a system that is used in mobile devices.

5.4.2 Using Threads in Mobile Java

Using threads in Java is simple. There is a type `thread`, that can be instantiated as a `Runnable`, which then creates a new thread. The new thread starts its execution from method `run` after the creator calls the `start` method of the thread. Mutual exclusion is implemented either in the fashion the methods are called or using the `synchronized` keyword, which enables monitor-like implementations.

As an example, consider the following case. There is a shared variable that is shared by two classes. The variable is hosted by `IncrementerThread`, but the other thread (`MyThread`) is allowed to access the value directly.

The hosting thread (instance of `IncrementerThread`) will only increment the shared integer value. The definition of the thread is the following:

```
public class IncrementerThread
    extends Thread {
    public int i;
    public IncrementerThread() {
        i = 0;
        Thread t = new Thread(this);
        t.start();
    }
    public void run() {  for(;;) i++; }
}
```

The other thread runs a midlet that acts as the user interface for accessing the shared variable. Operations are given for exiting the program and for addressing the shared variable. Moreover, the midlet owns the thread that hosts the shared value:

```
import javax.microedition.midlet.*;
import javax.microedition.lcdui.*;

public class MyThread
    extends MIDlet
    implements CommandListener {

    private Command exitCommand, toggleCommand;

    private TextBox tb;
    private IncrementerThread t;
```

The constructor of the midlet simply creates new commands to be used for controlling the execution:

```
    public MyThread() {
        exitCommand = new Command("EXIT", Command.EXIT, 1);
        toggleCommand = new Command("Toggle", Command.OK, 2);

        t = null; // No thread is instantiated in this phase.
    }
```

The normal operations associated with the method are the obvious ones, listed as follows. The command handler of the midlet allows the user to select exit or reading the shared value. The latter is managed in full in method `toggleThread`:

```
protected void startApp() {
    tb = new TextBox("Thread Example",
                     "Thread was started", 80, 0);
    tb.addCommand(toggleCommand);
    tb.addCommand(exitCommand);
    tb.setCommandListener(this);
    Display.getDisplay(this).setCurrent(tb);
    t = new IncrementerThread();
}

public void commandAction(Command c, Displayable d) {
    if (c == exitCommand) {
        destroyApp(false);
        notifyDestroyed();
    } else { toggleThread(); t.interrupt(); }
}

protected void destroyApp(boolean u) {}

protected void pauseApp() {}
```

Method `toggleThread` is responsible for the main actions of the midlet. When called, it reports the value of the shared variable to the user, again using a `TextBox`:

```
private void toggleThread()
{
    tb = new TextBox("Thread Example",
                     "Thread run " + t.i + " times",
                     80, 0);
    tb.addCommand(toggleCommand);
    tb.addCommand(exitCommand);
    tb.setCommandListener(this);
    Display.getDisplay(this).setCurrent(tb);

} // toggleThread
} // MyThread
```

The application can now be run to verify the behavior of the tasking system. Toggling the thread with the associated button gives an idea on how efficiently threading takes place when run by the underlying virtual machine.

5.4.3   Problems with Java Threading

Perhaps the main problem of the scheme presented above is that changing between threads is a very costly operation. As it is the virtual machine that handles both threads, and as the hosting operating systems can consider the virtual machine as a separate system from an implementation perspective, to which only restricted resources are given in the first place, forcing the virtual machine to perform excessive threading is harmful for performance. As a result, the implementation leaks and becomes visible to the developer.

Another issue worth considering is that the threads are rather independent entities, which is in line with the principle that objects are entities of their own. For instance, in order to introduce an operation that would terminate the thread in the above system, commonly considered implementation alternatives where a master thread would directly control the life of an auxiliary thread do not work. Firstly, simply setting the thread to `null` when it should terminate will not lead to the garbage collector eliminating the thread, although this could be considered a rational interpretation, as there are several references to it in the internal representation of the run-time environment. Furthermore, based on studying practical implementations, operations for terminating threads have been left out from early versions of MIDP Java, and even the versions that do support it do not work too well with such a tight loop as used above. Therefore, the correct way to terminate the thread would be to extend the incrementer loop in a fashion that would allow the thread to decide to terminate itself. For example, the following code could be used:

```
public void run() {
    while (still_running) i++;
}
```

Then, stopping the thread would only require setting variable `still_running` to false in some other method of the class. However, continuous checking of this condition is an obvious overhead of this solution that would be eliminated if thread management were implemented in a different fashion.

5.5   Symbian OS and Concurrency

As in many other respects as well, the Symbian OS concurrency mechanisms rely on the talent of the designers. On the one hand, one can use common facilities, such as threads and processes, which are available in Symbian OS. On the other hand, a framework called active objects is included that enables simplified, pseudo-parallel executions that use less resources. However, the correct use of the framework requires that the developer introduce an adequate specialization of the framework.

5.5.1   Overview

Symbian OS threads are similar to those that can be commonly found from many operating systems, and they can also be used in a similar fashion. However, in many cases, it is more adequate to use active objects, which is the Symbian OS way to perform multitasking in a resource-friendly fashion. In addition, the normal infrastructure for concurrency, including for instance semaphores, is included in the platform.

In the following, we give an overview to using threads and active objects. Towards the end of the section, the focus is placed on the latter, as threading in the Symbian OS environment is not too different from other platforms on the one hand, and active objects are the fashion the platform favors at the level of application on the other hand, taking practicalities of the restricted environment into account.

Threads

As such, Symbian OS threads, implemented as class `RThread`, are similar to the common thread concept already discussed earlier. However, at the level of details, there have been several upgrades when the operating system has been incorporated with enhanced real-time properties in Symbian OS v.8.0B.

Let us next consider a sample program that demonstrates the use of threads in the Symbian environment using a shared variable i to verify the effect of executions. The structure of the example is similar to that of the above Java example, except that instead of classes, we simply use procedures that provide the means to execute threads in the Symbian environment. Here, variable i is used for communication, and `myOp` is the operation that is repeatedly executed by the newly created thread. Moreover, explicit `kill` is used to terminate the thread, a detail that was not an option with mobile Java.

First, we need some header files that define the necessary thread and Symbian infrastructure. In addition, we introduce the shared variable:

```
#include <e32cons.h>
#include <e32std.h>

TInt i; // Communication variable.
```

For every new thread, a function is needed that will act as the main procedure of the thread, referred to as thread function. In this example, we use a simplistic function `myOp` that increments the value of the shared variable i in a continuous fashion:

```
void myOp(TAny* aArg) // Procedure run by a new thread.
    {
    for(;;) i++;
    }
```

Next, we give operation `MyConsoleL` which is responsible for the management of the new thread. First, the operation resets the shared variable, then it creates a new thread that increments the value of the shared variable, and when the user hits the keyboard, the operation terminates the new thread and reports to the user the number of executed loops in the new thread. This is implemented as follows:

```
void MyConsoleL()
    {
    LOCAL_D CConsoleBase* console;
    RThread myThread;

    console = Console::NewL(_L("Thread-use"),
                            TSize(KConsFullScreen,
                            KConsFullScreen));
    CleanupStack::PushL(console);

    i = 0; // Shared thread is reset in the beginning.

    TInt status = // Create thread.
        myThread.Create(
            _L("TEST THREAD"), // Thread name.
            (TThreadFunction)myOp, // Called procedure.
            0x1000, // Size of stack.
            0x1000, // Minimum size of thread's heap.
            0x1000, // Maximum size of thread's heap.
            NULL, // Data to be passed to the thread.
            EOwnerProcess); // Owned by the current process.

    User::LeaveIfError(status);
    // Continue only if thread creation was successful.

    console->Printf(_L("Thread created.\n"));

    myThread.Resume(); // Activates the thread.

    console->Getch(); // Wait for keyboard hit.

    myThread.Kill(KErrNone); // Terminate the thread.
    myThread.Close(); // Closes the connection to the thread.

    console->Printf(_L("Thread run %d rounds.\n"), i);
    console->Getch(); // Wait for user action.

    CleanupStack::Pop(); // Console
    delete console;
    console = 0;
    }
```

Finally, the main program is needed that creates the cleanup stack of the application and calls `MyConsoleL`. This is introduced in the following:

```
GLDEF_C TInt E32Main() // main function called by E32
    {
    __UHEAP_MARK;
    CTrapCleanup* cleanup=CTrapCleanup::New();

    TRAPD(error, MyConsoleL());
    __ASSERT_ALWAYS(!error,User::Panic(_L("EPOC32EX"),error));

    delete cleanup;
    __UHEAP_MARKEND;
    return 0;
    }
```

Once created, Symbian OS threads are pre-emptively scheduled, although kernel threads can also be scheduled in a non-pre-emptive fashion to enable improved real-time features. The currently running thread is the highest priority thread that is ready to be run. When several threads have the same priority, they get execution time in slices using the round-robin algorithm. Obviously, in a multi-threaded application, thread priorities must be carefully designed. When needed, the priority can be set using `RThread::SetPriority` method.

In addition to managing priorities, a thread can be suspended (operation `Suspend`) and resumed (`Resume`), which was already introduced in the above example, as well as terminated (methods `Kill` and `Terminate`). Obviously, a thread can also be terminated with a panic, which is to be reserved for special events. One can also register to obtain a notification when a thread is terminated (`RThread::Logon`), and query the reason for the termination (`RThread::ExitType`).

As threads are often located in different processes, they cannot communicate via shared memory. Instead, inter-thread data transfer is to be used. Methods called `ReadL`, `WriteL`, `GetDesLength`, and `GetDesMaxLength` have been provided for this purpose. The use of these methods is however dependent on the Symbian OS version used, as the implementation in the kernel has been altered.

Active Objects

Active objects are a Symbian-specific way to implement concurrent executions in a serialized fashion. The rationale is that by using a serialized implementation, where one thread serves several active objects rather than using a separate listener for every possible incoming event (Figure 5.1), the following advances can be gained.

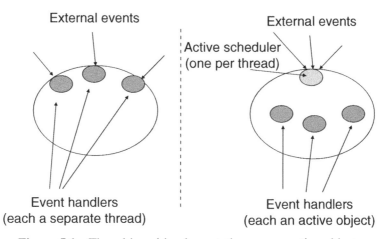

Figure 5.1   Thread-based implementation versus active objects

To contrast the use of threads, we will introduce a simple active object in the following. In general, the use of active objects is based on the following concepts:

1. *Active object*, which includes the operations that should be performed in a concurrent (or more precisely, pseudo-parallel) fashion.
2. *Active scheduler*, which is the object responsible for scheduling active objects into execution.
3. *Service providers*, which provide services to active objects and trigger their executions in a parallel fashion. For instance, they may communicate with external systems.

Figure 5.2 illustrates the use of active objects. The different phases introduced in the figure have the following semantics:

1. An active scheduler, i.e., a software component responsible for active objects' scheduling, is created and installed in the Symbian OS infrastructure. A thread is always associated with the scheduler. In code, class `CActiveScheduler` or its derivative must be used.
2. Active object, derived from `CActive`, that is needed is created and added to the scheduler. In addition, a request is made for a service.
3. Active object's request function activates the service provider. Then, the active object sets its `iActive` member variable to true to signal that it is now ready to serve incoming events, and `iStatus` to pending to signal that a request is now pending.
4. Scheduler is started, and the active object now waits for incoming requests.

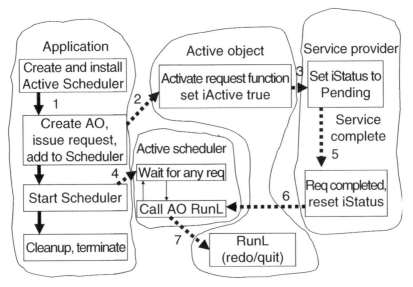

Figure 5.2 Operation of active objects

5. When service is completed, iStatus is reset. This will be signaled to the active
   scheduler.
6. When active scheduler is activated, it calls active object's RunL method.
7. When RunL completes, it can either quit or redo the service request.
8. When the execution of the system completes, all objects are deallocated.

In a nutshell, a single thread is used for pseudo-parallel treatment of a number
of operations, which are defined using active objects. If necessary, it is possible to
define a derived version of the active scheduler that includes application-specific features. The implementation is wrapped in a framework in accordance to the principles
of object-oriented design. Unfortunately, due to the simplicity of the implementation and the complexity of the design, capturing the behavior at the level of
implementation is often easier.

Discussion

To summarize, in Symbian OS, the special value of threads is in being able to act
as the unit of execution. However, actual concurrent executions are usually handled
with active objects that offer several benefits.

- Design is eased, as all operations are executed in a non-pre-emptive fashion. This
  liberates the developer from considering mutual exclusion.
- Memory can be saved. The memory consumption of an active object is some
  bytes, whereas a thread needs 4 kB of kernel memory, 12 kB of user-side

memory. Some memory is required for hosting an object of class DThread inside the kernel.

- Performance is improved for the total system, as no context switching is needed, and the execution is not stopped in favor of other events served by the same thread.
- Generic event handler can be offered in application architecture or some other library component.

Let us next consider a practical implementation of an active object at the level of code.

5.5.2 Sample Active Object

The particular active object we will introduce next uses a timer as an external service that will asynchronously generate an event after a certain time (RTimer). In addition, to simplify the program we now give a global variable console, which is used in both the main program and the active object. Otherwise parameters would have to be used to inform the active object about the console to be used in RunL. This in turn might confuse the actual logic of the operation:

```
#include <e32base.h> // Needed for active objects.
#include <e32cons.h> // Needed for using constants.

LOCAL_D CConsoleBase* console; // Write all messages to this.
```

As usual when composing active objects, the class to be developed is derived from class CActive:

```
class CDelayer : public CActive
    {
```

Methods of the class include normal facilities for two-phase construction (constructor, ConstructL, and NewL), destructor, and three methods related to active object features. SetHello sets the timer to cause an event after a delay given as a parameter, RunL is the method that contains the actual operation, and DoCancel is used to cancel waiting for an event. Methods RunL and DoCancel belong to the methods of CActive. At the level of code, this results in the following definition:

```
public:
    static CDelayer *NewL();
    ~CDelayer();
    void SetHello(TInt aDelay);
private:
    CDelayer();
    void ConstructL();
protected:
```

```
void RunL();
void DoCancel();
```

In addition, the class includes a timer that is used as the event generator. This is implemented as a private variable, as it is only used by the class itself:

```
private:
    RTimer iTimer;
    }; // class
```

This completes the header file. Next, we introduce a sample implementation for the active object. To begin with, we introduce the normal constructors and destructor:

```
CDelayer::CDelayer(): CActive(EPriorityStandard) {}
// Priority of the active object is EPriorityStandard.

void CDelayer::ConstructL()
    {
    User::LeaveIfError(iTimer.CreateLocal());
    CActiveScheduler::Add(this);
    // Add this object to the set of active objects.
    }

CDelayer * CDelayer::NewL()
    {
    CDelayer * self = new (ELeave) CDelayer();
    CleanupStack::PushL(self);
    self->ConstructL();
    CleanupStack::Pop(); // self
    return self;
    }
CDelayer::~CDelayer()
    {
    Cancel();
    // DoCancel implementation requires that destructor calls
    // Cancel method.
    iTimer.Close();
    }
```

In addition, a method is given for activating a service from some other active party. In this particular implementation, we use iTimer, which is requested to cause an event after aDelay microseconds. Finally, when the request is completed, this active object begins to wait for the event (SetActive):

```
void CDelayer::SetHello(TInt aDelay)
    {
    // Nested set operations are forbidden. This is ensured
    // by the following assert.
```

```
__ASSERT_ALWAYS(!IsActive(),
            User::Panic(_L("CDelayedHello"),
                        KErrGeneral));
// Request for an event after aDelay.
// iStatus is a member variable of CActive.
iTimer.After(iStatus, aDelay);
// Set this active object active.
SetActive();
}
```

Finally, the operation that takes care of the actual task is needed. In this sample implementation, RunL writes some text to the console, and stops the active scheduler. This returns the control to the point where the scheduler was started:

```
void CDelayer::RunL()
    {
    _LIT(KTimer, "\nTimer expired");
    console->Printf(KTimer);
    // Stop the active scheduler, so the execution can continue.
    CActiveScheduler::Stop();
    }
```

Statement CActiveScheduler::Stop stops the execution of this active object. As a result, the control returns to the main program. If needed, it is possible to perform something totally different. Furthermore, if the active object introduced by Symbian application architecture is used, this should not be included in user code in the first place, assuming that the application relies on the use of a single thread. For multi-threaded applications, it is not uncommon that one has to implement one's own active scheduler.

In addition, the DoCancel operation must be provided in order to cancel any outstanding request. In this particular case, it is enough to call the cancel method of the associated timer:

```
void CDelayer::DoCancel()
    {
    iTimer.Cancel();
    }
```

Next, we instantiate the active scheduler and the console into which the message is sent. This is implemented in operation MyConsoleL listed in the following. Note that console is now a global variable, not an automatic one, and therefore it is not pushed to the cleanup stack:

```
int MyConsoleL()
    {
    console = Console::NewL(_L("Active Object Sample"),
                            TSize(KConsFullScreen,
```

```
                                        KConsFullScreen));

   console->Printf(_L("Creating Scheduler\n"));

   CActiveScheduler * as = new (ELeave) CActiveScheduler();
   CleanupStack::PushL(as);
   CActiveScheduler::Install(as);

   console->Printf(_L("Starting Scheduler\n"));

   CDelayer * d = CDelayer::NewL();
   CleanupStack::PushL(d);
   d->SetHello(3000000); // Time in microseconds.

   CActiveScheduler::Start();

   console->Printf(_L("Ready.\n"));
   console->Getch();

   CleanupStack::Pop(2); // d, as
   delete d;
   delete as;

   delete console;

   return 0;
   }
```

In this code snippet, the call to method `CActiveScheduler::Start` plays an important role. It moves the control to the active scheduler, which starts to wait for events from the environment. In this particular case, the environment consists of the timer introduced above. Again, because this particular program is single-threaded, this line would not be needed if Symbian application architecture was used, but it would be enough to exit from the program.

Finally, we introduce a main program that introduces functions similar to those already introduced in previous chapters:

```
GLDEF_C TInt E32Main()
    {
    __UHEAP_MARK;
    CTrapCleanup* cleanup = CTrapCleanup::New();

    TRAPD(error, MyConsoleL());
    __ASSERT_ALWAYS(!error,User::Panic(_L("EPOC32EX"),error));
```

```
    delete cleanup;
    __UHEAP_MARKEND;
    return 0;
    }
```

This operation completes the sample implementation, and it is now ready for compilation and test runs.

5.5.3 Active Objects and Applications

As already discussed, Symbian applications obeying the structure introduced in Chapter 3 always include an active scheduler. This facility is the event handler inside the application architecture that was already used for receiving events from the graphical user interface.[1] Thus, the extension we introduce here can be considered as an implementation mechanism of the event handling scheme. In terms of an implementation, user interface components are actually generating events that the active object of the application automatically handles by dispatching the control to the right component.

A common design for the implementation of active objects and applications' graphical user interfaces is the use of the Observer pattern to implement the necessary callbacks. Technically, one introduces an auxiliary interface, say `MObserverIF`, that is implemented by the view or the controller of an application. Then, when defining an active object, a reference to this interface is enough, and only when instantiating an actual object, the actual reference to a concrete element is needed, following the normal callback routine.

5.5.4 Problems with Active Objects

Like any design solution, active objects are not without problems. One obvious problem is what to do when an error occurs during the execution of `RunL`. This has been solved by allowing method `RunError`, whose role is to handle such exceptions elegantly whenever possible. However, it is not obligatory to implement the method, but a default implementation has been provided that can be overridden when defining the active object. The default implementation returns an error code to the object's active scheduler, which then reacts to this code by panicking the application. Therefore, if recovery actions are desired, a corresponding implementation should be given.

Another problem follows from the fact that the execution of `RunL` cannot be interrupted. Therefore, it is possible that a higher-priority event handling must wait until a lower-priority event has been handled in full. As a solution, long-running

---

[1] More precisely, from the so-called Window Server. We will address Symbian servers in more detail in the following chapter.

RunL must be split into several parts and be replaced by a stateful design that knows from where to continue the next time the method is called. While such problems are common in event-driven programming, this obviously increases the complexity of the design. Using State design pattern introduced by Gamma et al. (1995) can result in improved readability of the final system.

Finally, although active objects are only a mechanism for implementing event handlers, the implementation can be considered relatively complex. Problems in for instance splitting RunL can result in difficulties later on when new active objects are introduced in the scheduler even if the original active objects remain unaltered. This kind of error can be easily introduced as the system evolves and more tasks are integrated to it. Furthermore, properties of object-orientation are extensively used, and a programmer who is not fluent in object-oriented design can face difficulties in using the scheme, and would value a more traditional implementation setup. In other words, the downside is that the framework does not really introduce an abstraction that could leak as such, but rather an implementation architecture to which the programmer must adapt. In particular, the task of a programmer is not eased considerably when considering an implementation of an explicit event handler, in contrast to active object abstraction that gains its properties from the underlying implementation on the one hand, and from capable developers on the other.

5.6   Summary

- Concurrent programming can lead to complex designs regarding critical regions. Furthermore, debugging is often difficult.
- Using program-level pseudo-parallel executions instead of more elaborated concurrency schemes can offer benefits regarding:

  1. improved portability,
  2. smaller memory footprint,
  3. simplified programming model and eased debugging.

- Supporting built-in facilities has been included in programming environments.

  - Mobile Java can offer thread simulation inside the virtual machine, which eases porting and lets the programmer act as if real threads were used.
  - Symbian OS encourages the use of active objects, which must be taken into account in full by the programmers.

- Observer pattern can be used for enabling the communication between different parties.

5.7 Exercises

1. What kind of access should a scheduler provide to its thread management in order to allow a Java virtual machine implementation that relies on operating system threads instead of green threads? What kinds of data structures would then be needed in the virtual machine?

2. Compare MIDP Java's and Symbian OS's implementation of pseudo-parallel executions. what differences must the application developer consider when using them?

3. Sketch an architecture for a system where the benefits of the Symbian OS active object scheme would be received, using a design that would more explicitly show that the system is actually an event handler.

4. Implement a number of sample active objects that listen to several events (for instance a number of timers). Does the number of parallel (or almost parallel) events become visible in the execution? How would the developer be instructed to use the system?

5. What types of properties make active object abstraction leak? How about Java's threads that are emulated inside a virtual machine?

6. What kind of a design would allow pre-emptive active objects? What kinds of advantages and downsides would this imply?

7. Implement the corresponding extensions to the Symbian OS threading example as was introduced for Java, i.e., a class encapsulating the shared variable. Then compare the performance of threading of Java and Symbian OS. Add also additional threads and implement mutual exclusion in terms of `Synchronize` operations and semaphores. How will these affect the comparison?

6

Managing Resources

6.1 Resource-Related Concerns in Mobile Devices

A mobile device is a specialized piece of hardware. It has several different types of resources that may require special monitoring and management activities that require scalable yet uniform designs. In addition to hardware resources, when the level of abstraction is raised, some parts of software can be treated as resources as well. At the very least, applying similar techniques to them can ease the development.

Based on the above, resource management is an important concern when aiming at software running in a mobile device. Following the conventional guidelines of modularity addressed by Parnas (1972) and Parnas et al. (1985) for instance, implementing the access to resources is therefore a good candidate for separation from the rest of the system. In this section, we discuss the different paradigms of separation and their implications.

6.1.1 Overview

In many ways, each resource included in a mobile device can be considered as a potential variation and management point. Additional motivation for this is gained from the fact that different configurations used in actual phones seldom follow the same line of hardware, but include updates due to pricing and logistics reasons, for instance. Moreover, many resources offer different facilities to higher levels of abstraction, Therefore, it is only natural that the underlying software architecture should adopt a principle where each resource is associated with a manageable, dedicated software entity. Predicting the actual upgrades of individual pieces of hardware can be difficult but, in general, it is obvious that hardware updates will be encountered.

In this kind of situation, the most obvious strategy is to embed hardware dependencies in modules (Figure 6.1). For instance, the parts that are associated with the file system and disks in general should form a subsystem. In other words, we are

Programming Mobile Devices: An Introduction for Practitioners Tommi Mikkonen
© 2007 John Wiley & Sons, Ltd

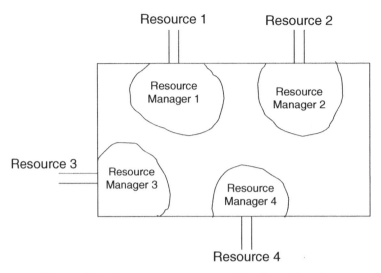

Figure 6.1   Resource managers and hardware interface

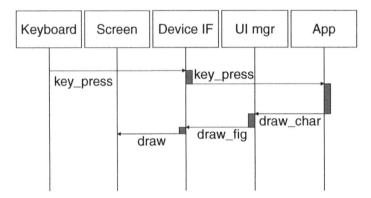

Figure 6.2   Sequence of events following a key press

essentially creating a manager for all aspects of the file system inside a specialized entity. This gives a clear strategy for connecting software to the underlying
hardware. First, a device driver addresses the hardware, and on top of the driver,
a separate resource manager takes care of higher-level concerns. Subsystems can
communicate with each other by for instance sending messages to each other. As
an example, Figure 6.2 represents a situation where the user hits a keyboard, which
triggers a sequence of operations in different resource managing modules. A practical rule of thumb then is that introducing more management and control functions
inside modules gives better predictability, but consumes resources for management
tasks that do not serve the fundamental purpose of the device.

In order to isolate such functions or generic service providers from the rest of the system, the normal facilities of an operating system can be used. Process boundaries can be used for separating the different resources at the level of an implementation. Unfortunately this essentially makes the option of handling errors that occur in the manager of a certain resource more difficult. Still, because it is unlikely that a special-purpose client application using a resource would be able to make the resource manager recover from the error, it is possible to embed also recovery instructions inside the resource manager. Another problem in isolating resource managers is memory protection: in many cases resource managers can use the same data but memory protection may require the use of copies. A practical guideline for designing such isolation is that it should be possible to reconstruct all events for debugging purposes. This allows one to study the causes of different sequences of events in isolation and to reveal errors associated with cases that are seldom executed in practical implementations.

There are two fundamentally different solutions for embedding resources in the system. The first solution is to put all resources under one control. This can be implemented using a monolithic kernel or a virtual machine through which the access to the resources of the device is provided. The alternative is to use an approach where all resource managers run in different processes and the kernel only has minimal scheduling and interrupt handling responsibility (microkernel). In the following, we discuss these alternatives in more detail.

6.1.2 Grouping Resource Managers

A monolithic design, where several, if not all, resource-related operations are embedded in the OS kernel, requires a design where the kernel includes a large amount of code and auxiliary data. While these facilities, such as the file system, need not be part of the kernel for technical reasons, it is often beneficial to include them in the kernel, as in many cases they share data structures or require communication with other data structures included in the kernel. The situation is illustrated in Figure 6.3, where ellipses denote resources and application processes, and the monolithic kernel is shown as a rectangle. The interface to the kernel can be understood as an API to all the resources that are accessed via the kernel, although in a practical implementation an interrupt-like routine is used. A practical example of such a system is Linux, where the kernel is in principle monolithic, but dedicated modules are used for several rather independent tasks, like processor and cache control, memory management, networking stacks, and device and I/O interfaces, to name some examples.

Such a system is commonly implemented in terms of (procedural) interfaces between resources. Unlike a user call to these operations discussed above – when applicable in the first place – which leads to an interrupt, kernel code can usually call different routines directly. In addition, as data structures are often shared, adequate mechanisms must be used to ensure correctly implemented mutual exclusion.

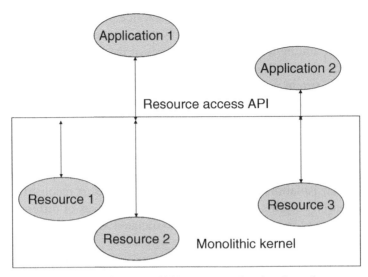

Figure 6.3   Monolithic resource-hosting kernel

Positive and negative aspects of this approach are the following. Addressing different parts of the kernel using procedural interfaces can be implemented in a performance-effective fashion as no context switching is required but all the resources can be accessed directly. Then, the operating system can serve the requests of programs faster, in particular when an operation that requires coordination in several resources is needed. On the downside, without careful management, it is possible to create a tangled code base in the kernel, where the different parts of the system are very tightly coupled. In practice, an additional detail that seems to complicate the design of a monolithic kernel is that it often seems practical to add some more properties to it, since there are no explicit boundaries on what to include in the kernel and what to leave out. Over an extended period of time, this can make the kernel harder to manage.

6.1.3 Separating Resource Managers

Parallel partitioning of resource-management-related facilities of a mobile device leads to a design where individual resources are managed by separate software modules. These modules can then communicate with each other using messages, leading to an architecture commonly referred to as *message passing architecture* (Shaw and Garlan 1996).

The microkernel approach can be considered as a message passing system where the operating system's kernel is used as a message passing bus, similarly to the approach of Oki et al. (1993), which introduced the idea in the context of distributed systems. Then, following the scheme of distributed systems, different functions are separated to different processes, and the messages used for communication with

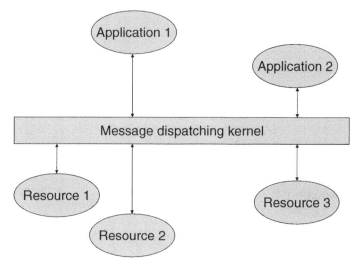

Figure 6.4 Message dispatching kernel

resources form the interface for accessing resources. The approach is illustrated in Figure 6.4. Obviously, it is possible to hide actual messages using for instance dynamically linked libraries that offer a procedural interface rather than messages, whose practical use can be considered more complex.

One common implementation for this type of architecture is that all the modules run in a process of their own. Inside a module, a thread is used to listen to messages. Whenever a message is received, the thread executes a message-handling procedure, which fundamentally is about event processing. While this design decision adds flexibility to the system – new modules can be safely introduced, if they rely on the use of new messages and in particular do not send messages that can affect the behavior of already existing parts of the system – it also has its downsides. The principal problem is that in many cases, communication needed for passing the messages requires several context switches, if only one processor is available; first, a process hosting the resource sends a message, then the kernel dispatches it to the recipient(s), and finally the message is received by another process. In particular, operations that require complex cooperation of several resources easily become expensive to perform. Therefore, while a design where all resources are located in different processes is beneficial for separating concerns at design level, in addition to plain message passing other means of communication, such as shared memory, are usually enabled, at least in some restricted form (Figure 6.5). A more sophisticated design can be composed when some of the resources are run in different threads but within the same process. Then, they automatically share the same memory space, and there is no need for special operations for accessing shared memory.

Even if a monolithic design is used when considering the kernel, it is possible to manage resources implemented with software using the principles of message

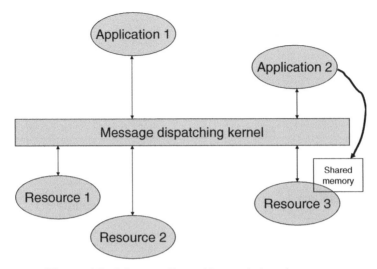

Figure 6.5   Message dispatching and shared memory

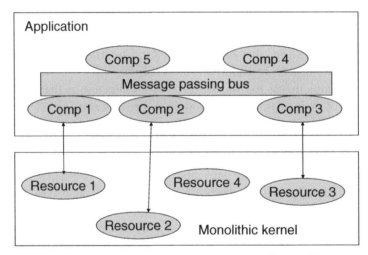

Figure 6.6   Application and message passing bus architecture

passing architecture (Figure 6.6). For instance, a design can be used where message passing architecture is used on top of the kernel and the benefits of message passing architecture are gained only at the level of the application. Analogously to system design discussed above, also at the level of application it is possible to encapsulate properties associated with resources in one component.

When considering implementations of mobile device platforms, also message passing based architectures have been used. Perhaps most notably Symbian OS

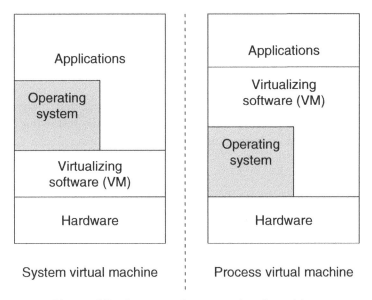

Figure 6.7   System and process virtual machines

builds on this type of system despite the considerable overhead introduced due to inter-process communication.

6.1.4   Resource-Hosting Virtual Machine

One more approach is to introduce a separate software entity, a virtual machine to host resources of the underlying platform. Virtual machines can be of several levels of abstraction, ranging from low-level virtual machines that can be considered a feature of the underlying hardware to complete interpreters that can offer sophisticated services, and their features may vary accordingly. As a result, virtualization can take place in different places of the system, in particular in between applications and hardware and in between applications and infrastructural software (Smith and Nair 2005). These approaches, referred to as system and process virtual machines, are illustrated in Figure 6.7.

The benefits of using a virtual machine in this context are the expected ones. Firstly, porting can be eased, and in fact it is possible to define a standard execution environment, with well-defined, standardized interfaces, as defined by mobile Java. Moreover, techniques such as dynamic compilation can also be used.

On the downside, performance loss commonly associated with virtual machines is worsened in an environment that is constrained in terms of resources. Therefore, many devices include native software to implement main routines. However, in practice nothing forbids implementing a virtual machine based mobile device, where the majority of routines would be implemented with translated (or interpreted) techniques, and in fact a number of such devices have been implemented, building for instance on the use of Java.

6.2 Common Concerns

No matter which approach to handling local resources is selected, there are multiple concerns to be taken into account. The approach taken in the following is to address the concerns listed below in more detailed discussion.

6.2.1 Overview

There are several common concerns when considering resource use of a mobile device. Many of them are derivatives of scarce hardware resources, but some of them can be traced to the requirements of organizations developing mobile systems. For instance, one can consider the following concerns:

- *Extension and adaptation* is needed for being able to reuse the same code base in different devices and contexts whose hardware and software characteristics and available resources may differ. For instance, some devices can use hardware-supported graphics acceleration, whereas others use the main processor for this task.
- *Performance* requires special attention, because in many cases mobile device hardware is slower than workstation on the one hand, and, due to the generic nature of mobile devices as application platforms, optimization for a single purpose is harder than in a purely embedded setting on the other hand.
- *Energy management* is an issue that arises when mobile devices are used more like computers and less like single-purpose devices; being active consumes more memory.
- *Internal resource management* is needed for ensuring that the right resources are available at the right time. Furthermore, it may be a necessity to introduce features for handling hardware-specific properties. The issue is emphasized by the fact that in many cases, resource management is always running in the background.

In the following, we discuss some implementation techniques for these concerns.

6.2.2 Extension and Adaptation

Probably the most well-established approach to extensions and adaptations is to implement common parts in one component, and to isolate all variance to different modules. Based on the ideology of software product lines (Bosch 2000; Clements and Northrop 2002), the goal is to decompose a system into parts so that different systems can be built using the same parts. Moreover, parts that have already been implemented should not be wasted but they should be reused in other products as well.

In mobile devices, the most obvious source of resource-motivated adaptation for mobile devices is the hardware that is used. There are screens of different sizes, the amount of memory can be altered, the keyboard can be replaced with a touch screen, and so forth. Less obvious but still related to hardware are smaller changes that may be invisible to the user of the phone, but require changes in the software implementation. An example of such an event is an upgrade of camera to another whose resolution is the same, but whose implementation is upgraded by for instance enabling more liberal control over power management. Furthermore, when using a common operating system, hardware or device manufacturers may need to adapt the operating system to the specifics of the hardware.

In addition to hardware-level variation, also features implemented with software may vary. For instance, market segmentation and targeting of mobile devices to different types of users can result in a situation where a number of features are removed (or replaced with some other features). While this is not a resource in the strict sense, similar techniques can be applied.

A commonly used technique that lends itself to extension and adaptation is the use of plugin components, as already discussed in Chapter 4. In the context of resources and their management, one can use plugins for the specialization of the system to fit a certain environment. When comparing device drivers and plugins, device drivers are a considerably lower-level technique, whereas plugin components can be used for any implementation. In some cases plugin components can be extended, specialized or adapted with further plugins, or plugin wrappers can be used as adapters for different types of resources. The downside of using plugins is that their instantiation impairs performance. However, this need not be a problem, as plugins can be specialized for a particular piece of hardware that will be used in the target device.

Another way to perform extension and adaptation is to use aspect-oriented software development, where new types of facilities are offered with which it is possible to localize types of things that commonly need to be tangled in code (Filman et al. 2005). For instance, logging is a commonly used example, but also other types of features that need consideration in several modules can benefit from aspect-orientation. In the scope of mobile devices, probably the most prominent implementation approaches are AspectC++ (Spinczyk et al. 2002) and AspectJ (Miles 2004), which extend C++ and Java with an option to weave aspects, i.e., pieces of code that know where they belong in the code, into so-called pointcuts that for example refer to method calls of a certain type, to completed C++ and Java programs. Because aspects essentially enable an approach where a baseline implementation is given with standard C++ or Java and additional features are added with aspects, it is tempting to first focus on the application behavior in the baseline. Then, additional concerns, such as creating different variants or including energy management, following the spirit of Sadjadi et al. (2002), or even generic resource management features, can be implemented with aspects on top of the baseline implementation. As the result, the baseline implementation would not be polluted with additional concerns, and the concerns would be clearly modularized. On the downside, an experiment on using AspectC++ as a tool for implementing variance has shown that in addition to tool chain problems in the interoperability of Symbian compilers and AspectC++, there are also problems with memory usage; in the experiment the size of a DLL almost doubled, which can be considered unacceptable for large-scale use in the mobile setting (Pesonen et al. 2006).

In addition, the structure of an application is an important source of adaptability and extension. Being able to separate unrelated concerns to different modules has been the goal of the Model-View-Controller pattern already introduced in Chapter 3. Separating the application logic to the model enables its reuse in different settings. Views, however, can be reused only if the properties of the different displays are relatively similar, and controllers only if the accessible controls are the same. Obviously, adding a software layer that manages this can be used, which adds complexity to software but eases reuse.

Finally, an additional issue is the freedom of selecting the desired features at the right time. For instance, one can consider conditional compilation, loading of dynamic libraries, and run-time adaptation as different alternatives. Obviously, the later the selection is, the more prepared the system must be to host the different features.

6.2.3 Performance

Resources are not always offering independent services to applications, but they may cooperate to jointly provide a more complex service. For instance, downloading a video stream from a file or network requires that several subsystems cooperate in an optimized fashion; first, radio hardware access is commonly implemented in a driver, then some protocol can be used for mapping the low-level data to a form that can be understood by applications, and finally an application is responsible for displaying the data. Moreover, several hardware resources may be used in the process to accelerate image processing or to control the transmission. How these subsystems then communicate becomes an issue, which is restricted by memory protection and physical implementation. Passing the data as such from one process to another can require extensive copying, which in turn can downgrade the performance. On the other hand, using shared memory that can be accessed by all subsystems can result in improved performance, but may be more error prone. In general, however, sharing memory rather than copying data several times can be considered a better option. Therefore, passing a reference rather than passing a new copy should be considered first, as this saves memory as well as reducing copying.

In addition, the different hardware facilities should be benefited from in full. This requires careful partitioning between the different processors, and must often be performed in application- and device-specific fashion, because the hardware configuration between different devices can vary considerably. Then, some devices may use software emulation for some features that others execute using a hardware accelerator. Finding a combination of design decisions that satisfy performance requirements in all different device configurations can be hard or even impossible. Again, iterative development can be used to allow prototyping, which can help in finding the worst performance bottlenecks in an early phase of the development.

A further topic regarding performance is that it is common that a device-specific implementation can be composed in a more efficient fashion than a generic one that can be adopted to different systems. However, the obvious price is that the implementation is restricted to a certain hosting device only.

Finally, in cases where there is no way to perform all the required tasks, a quality of service mechanism can be introduced. Then, even if the device is overloaded, the most important tasks can still be executed. Moreover, one should design applications so that failures are recovered from, and that applications can still continue in a degenerated form, following the guidelines of Noble and Weir (2001). For instance, if it is impossible to load a specific ring tone due to delays in file access for instance, the default ring tone is played.

6.2.4 Energy Management

Energy management is a common concern in a mobile device. While in the simplest case optimization can be considered as selecting the instruction sequence that consumes the least amount of energy, the gains are often minimal. Firstly, finding the right tricks for saving energy is platform specific. Secondly, even if an energy saving pattern is identified, applying it in practice can result in poorer results when considering energy consumption due to optimization, as pointed out by Surakka et al. (2005) and Tiwari et al. (1994), since compilers tend to optimize certain types of routines better than some others. For instance, even if it would be more memory efficient to use indices where a maximum number of bits is set, code generation can be so much better with indices starting from 0 that the energy saving does not pay off.

However, and often also more importantly, software plays an important role for managing other hardware elements than the processor. The ability to shut down and reinitialize pieces of hardware can then be used to manage energy consumption from the perspective of the whole device. In the simplest and most commonly used form, hardware is managed with timeouts. When a certain amount of time has passed, and a piece of hardware is not needed, it is shut down. Then, when the hardware is needed, it is restarted. As shutting down and restarting a piece of hardware takes more energy than just maintaining a piece of hardware active, the number of unnecessary shutdowns should be minimized without keeping the hardware active for too long. In practice, this can often be achieved through iteration. While timeouts seldom are an optimal solution for energy, they can often be considered as a good enough alternative, because they enable the hiding of energy management from higher-level applications. While applications' role in energy management can be considerable, timeouts, together with a rule of thumb that performance equals energy, are good rules of thumb at application level (Chakrapani et al. 2001; Seng and Tullsen 2003).

In addition to the straightforward option to turn a piece of hardware on and off, hardware can offer other, more elaborate, alternatives as well. For instance, processors can allow dynamic voltage scaling, where the frequency of the hardware can be tuned. Then, less energy can be used when there are fewer tasks to perform by using lower frequency, and when more performance is needed, the frequency can be raised at the cost of energy consumption. Another example is given by Pasricha et al. (2004), where an approach is proposed where dynamic adaptation is applied to background light to save energy. Furthermore, similar techniques can be applied to memory, where lower voltage maintains memory functional, but higher voltage is required for operations. Again, latency is associated with moving from lower voltage to higher voltage, which restricts the area of applicability of the technique.

While energy management is one of the most fundamental differences when comparing programming mobile devices to programming devices connected to fixed energy, for an application developer its role may be difficult to grasp. In fact, in many cases of applications, especially ones that are run for a short period of time and then shut down, one can neglect energy consumption, because the application is run only for a short period of time in any case. However, for the developers of low-level software, device drivers, and platform features, the issue is of crucial importance. Moreover, even with applications, one should rather allow turning the application off permanently when possible than to let it constantly occupy memory and use kernel resources.

6.2.5 Internal Resource Management

Managing resources consumes other resources. In particular, resource-related software requires memory to run in, and consume memory for storing the variables. As with all software, all the memory that is reserved remains under the control of the programs being executed or deallocated. For resource-related software, the issue is even more crucial than when implementing applications, because applications are usually shut down at least occasionally, whereas resource management remains active.

A common concern that must be considered when using any local resource is to determine whether or not a resource requires startup-time handling, 'Convenient time' handling, where the system determines when to activate a resource, or fully application-dependent handling, when the application has been activated. The approaches have different advantages and drawbacks.

- *Startup*. A piece of software that is continuously running consumes resources all the time, but when a resource whose management is based on this strategy is needed, the access is fast because no software startup is necessary. For instance, one could assume that a mobile device should constantly be ready for establishing an emergency call, so it may be advisable to reserve some critical resources beforehand. Another example of a task that is usually initialized at startup is the lowest-priority (or background) thread that is used for housekeeping and statistics. When the thread is scheduled for execution frequently, one knows that there are few real tasks to perform. In contrast, when the task is scheduled to execution seldom, the system is busy. It is also possible to allocate also other tasks to the thread, associated for instance with the management of other tasks or resource managers.

- *Convenient time management.* A resource that is handled by a convenient time management scheme can be activated by a timer, for instance. The approach is applicable in cases where one can make a prediction on how the user is going to behave. For instance, if the user decides to send an MMS, it may be applicable to consider if for instance the camera interface should be activated, just in case the user wants to take a picture. Similarly, even if the phone is switched off, it may be possible to start some software when it is plugged to electricity for the first time, thus making the first boot as fast as possible.

- *Run-time management.* Managing resources only when necessary more or less makes them application-specific, although they can be shared by several applications and offer their services in a centralized fashion. As a result, only resources that are necessarily needed are consumed. The cost of this approach is that in some cases the system can appear slow, when resources are handled during the initialization sequence of an application. On the other hand, it is usually enough to let the applications take care of managing the resources.

Sometimes it is necessary to make details of hardware visible in the interface used for accessing it. For instance, in Symbian OS there are two different types of operations for writing data to disk. One is such that the operation is considered completed when the data has been copied from the saving application's memory space to the memory space of the component that manages the disk, and the component is then responsible for writing the data to the disk; the other considers the disk write completed only when the data has been physically stored on the disk. As a result, the operation appears faster for the user in the first case, whereas the second alternative should be considered when the application wishes to ensure that the data is saved.

6.3 MIDP Java

From the viewpoint of a midlet and MIDP Java infrastructure, the hosting operating system appears as a monolithic service-offering kernel, no matter how it is really implemented. It offers all the services needed by Java, and cannot be directly influenced by it in any means apart from the interfaces.

In the following, we introduce the main principles of implementing a programming environment using such a strategy as the design guideline.

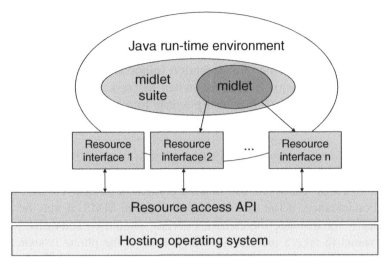

Figure 6.8   Access to resources from a MIDP Java application

6.3.1   Basic Elements

The basic elements of the MIDP Java infrastructure include the different interfaces that are offered for accessing resources of the device. From the perspective of the system, however, the environment acts as a single application, which has some predefined standard interfaces that can be used to access the resources of the device (Figure 6.8). In the following, we address the different types of resources that have been defined.

Interfaces to Resources

Access to resources in MIDP Java is based on interfaces that can be ported to different environments. Standards are defined using the Java Community Process (JCP), which enables any interested party to participate in the standardization process. Individual standards are referred to as JSRs (Java Standardization Request), which have been used to define also configurations and profiles mentioned earlier (Table 6.1). As an example MIDP v. 1.0 is JSR 37, defining:

- midlet application model,
- user interface libraries to access to the screen using simple graphical primitives,
- networking facilities for accessing external resources using HTTP,
- ability to use persistent memory for storing data using RMS (Record Management System), a simple record-based approach to storing data,
- timer support,
- system properties that will be discussed in the following,
- resource support for accessing files stored in the application's JAR file.

Table 6.1 Additional standardized resource-related facilities

JSR	Purpose
JSR 46	Foundation Profile
JSR 66	RMI Optional Package
JSR 75	PDA Optional Packages
JSR 80	USB API
JSR 82	Java APIs for Bluetooth
JSR 135	Mobile Media API
JSR 169	JDBC Optional Package for CDC/Foundation Profile
JSR 172	J2ME Web Services
JSR 177	Security and Trust Services API
JSR 179	Location API
JSR 180	SIP API
JSR 184	3D Graphics API
JSR 185	Java Technology for Wireless Industry
JSR 205	Wireless Messaging API 2.0
JSR 209	Advanced Graphics and User Interface Optional Package
JSR 226	Scalable Vector Graphics API
JSR 229	Payment API
JSR 230	Data Sync API
JSR 239	Java Binding for the OpenGL ES API
JSR 253	Mobile Telephony API
JSR 256	Mobile Sensor API
JSR 257	Contactless Communication API
JSR 258	Mobile User Interface Customization API
JSR 259	Ad Hoc Networking API
JSR 272	Mobile Broadcast Service API for Handheld Terminals
JSR 278	Resource Management API for Java ME
JSR 279	Service Connection API for Java ME
JSR 280	XML API for Java ME
JSR 281	IMS Services API

Similarly, MIDP v. 2.0 is JSR 118, which further introduces:

- improved user interface facilities that still remain compatible with v. 1.0,
- improved connectivity including HTTPS, datagrams, sockets, server sockets, and serial port,
- multimedia and gaming facilities including audio, tiled layers, and sprites,
- network push, i.e., opportunity to register midlets for activation when a device receives information from the server,
- improved security features,
- over-the-air provisioning (OTA).

In addition, both standards contain optional packages that can be included in an implementation, but this is not a necessity.

In addition to standard interfaces, it is possible for a device manufacturer to add new interfaces to enable new types of applications. Such interfaces are obviously a portability risk, as when the standard matures and extends, a standard solution could be implemented for the interface.

Interfaces to System Properties

In addition to the resources of the device, applications can get access to resource files using the special `getResourceAsStream` method. This allows applications to access resource files from their own context.

Also system properties (`java.lang.System.GetProperty`) can be studied. Available options include the following:

- `microedition.platform` returns the name of the device,
- `microedition.encoding` returns the encoding of characters,
- `microedition.configuration` returns the configuration and its version,
- `microedition.profiles` returns the profile and its version,
- `microedition.locale` returns the language and the country.

Another important aspect that has been introduced is the ability to register Java applications to listen to external events. The newer MIDP standard enables network-initiated midlets, which can be used for creating applications that are not dependent on user activity. This enables more sophisticated application.

To summarize, a clean and standardized software infrastructure has been created as a result of the above approach. However, there are some risks related to the different interfaces in different versions of standards as well as device manufacturer specific interfaces. Still, in principle the same applications can be run in different devices unaltered, as standard interfaces are required. However, in practice some problems have been encountered. We will address this issue in the following.

6.3.2 Compatibility between Different Devices

The fact that implementations are based on common standards has not cured the problem of variance in full. A number of features of applications may vary from one device to another. For instance the following facilities may be different in different devices:

- *Amount of memory*. Different devices may contain different amounts of memory, and they may give different amounts of memory for Java applications even if the actual device has more memory available.
- *Available interfaces*. Even if interfaces are standardized, older devices may include old proprietary implementations or lack interfaces completely. Moreover,

even if a certain resource is in principle available in a device's hardware, it may not be available for Java applications.

- *Implementation differences.* The same interface may have semantically varying implementations in different devices even if all of them claim to satisfy the standard.
- *Size and layout of screen.* Different purposes of devices result in different designs.
- *Keyboard layout.* Again, the purpose of the device can lead to different solutions. Moreover, facilities like a touch screen can be offered, which can result in different application design.
- *Standards.* Different versions of standards can result in compatibility problems.

A practical approach is to give developers some different device types, and have the first working version to run on all of them in a bearable fashion. Obviously, choosing these models such that their penetration is or is expected to be large can be considered desirable for practical reasons. An additional criterion for selection is that one should consider phones with different number and amounts of resources. For instance, using a very restricted phone in the beginning of the development can guide the developer to really consider resource consumption. A risk is that the resources of the low-end devices are really inadequate. For instance, if file access has physical restrictions that always imply a certain delay, it is impossible to fix this with software. However, another type of design might still be applicable, where using the disk is for instance hidden from the user.

Another aspect that should be considered is that MIDP Java applications are not in control of their own executions, but can be taken away even from execution state. As different devices may behave differently in some situations, this is yet another source of potential incompatibilities.

6.4 Symbian OS

As already mentioned above, the microkernel approach is used in the Symbian environment. This has resulted in the definition of special software components, whose purpose is to manage different types of resources. As is common in microkernel approaches, such resource managers are referred to as servers.

6.4.1 Servers as a Mechanism of Resource Management

In Symbian OS, specialized components called servers are used as the main elements of resource management. Every resource that the system has is encapsulated into a server, which is responsible for managing the resource. When the resource is used, the client first contacts the server, and establishes a session with the server. After establishing the session, the client can use resources and services offered by the server. Also error management is incorporated in the scheme. If a server is killed, error messages are given to the clients, and if a client dies, the server should be able to release allocated resources.

Table 6.2   Some Symbian servers and their responsibilities

Server	Purpose
File Server	Implements access to files.
Window Server	Implements Symbian OS's graphical user interface.
Database Server	Contains data that is shared by multiple applications.
Comms Server	Enables the use of serial port for different purposes.
Telephony Server	Implements telephony-related services for applications.
Messaging Server	Used to manage all message-based communication.
Camera Server	Access to camera and associated hardware.
Socket Server	Manages communications sockets and connections.
CONE Server	Control environment which is run in all applications.
Media Server	Enables audio and multimedia control.
Font and bitmap server	Manages all fonts and bitmaps.

In addition to plain hardware resource management, the use of servers is extended to other resources. For instance, some services of the Symbian kernel have been implemented in a server referred to as kernel server.[1] The server includes two special threads. One is kernel server thread, which is responsible for executing kernel functions. The other is null thread, which has the lowest priority, and therefore can be used for managing the processor's power consumption. This type of solution is common in many environments, as the lowest priority thread can easily perform housekeeping activities as well as measure the utilization rate of the device. Some sample servers have been listed in Table 6.2.

Naturally, servers can act as clients of other servers. This enables a layered architecture, where more primitive services of low-level servers are used by more abstract servers. A drawback of this approach is that there is a considerable overhead if servers communicate extensively. The reason is the underlying message dispatching architecture, where the kernel acts as the communication bus, and servers only communicate via it. As a solution, mechanisms have been offered for moving a block of memory from the memory space of one server to the memory space of another. Furthermore, sometimes servers run inside the same process, which

---

[1] The current Symbian OS Kernel and its main properties have been introduced in detail from a programming perspective by Sales (2005).

allows them to transmit data from one to another more liberally. For instance, communications-related servers have been implemented in such a fashion.

Even if servers are the prime mechanism for resource management, there are also other reasons to use servers instead of applications. In principle, if one wants to implement a subsystem that is being executed disregarding the execution of an application, a server must be implemented for such a purpose. In order to reduce the number of servers, different mechanisms have been implemented, which enable an external event, for instance an incoming SMS or WAP push message, to activate an application using so-called notifiers or publish-subscribe technique. The former is a Symbian OS standard mechanism to provide a user interface for a client thread; the latter is a technique introduced in Symbian v.9.0 that is based on so-called properties that correspond to system-wide variables, publishers being threads that define, set, and update properties, and subscribers threads that retrieve the values of properties and can also listen to changes in properties and react to them (Shackman 2005b).

Servers can be categorized in accordance to their life span. This leads to the following categories.

1. Server is initiated when the system is booted, and it runs until the system is shut down. This allows the system to perform operations disregarding the application.
2. Server is started when an application is run for the first time. However, when the application terminates, the server remains active. This enables systems where an application is used to activate some features in the system.
3. Server is started when an application is activated, and shut down when the activation terminates. This results in servers whose only intention is to offer services to certain types of applications. In some cases, implementation could be based on dynamically linked libraries as well, but the server approach is used for convenience.
4. Server is started when an application needs it. When serving is over, the server shuts itself down.

6.4.2 Implementing Servers

Server implementation is based on an active object that receives all the messages sent to the server. A class diagram depicting this is given in Figure 6.9.[2] As a result, each access to a resource can be considered as an event, and the natural event processing mechanism in Symbian OS is the concept of active objects.

However, unlike with active objects in general, servers are usually located in a process of their own. Their clients communicate with them with messages (RMessage). Messages are based on 32-bit-long identifiers, and in addition to the identifier they carry 4 parameters, each of which is 32 bits. Another important

---

[2] Exact classes used in the implementation vary slightly depending on Symbian OS version.

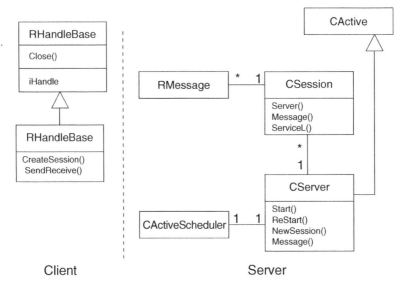

Figure 6.9 Active objects inside servers

concept is a session, which couples clients and servers. To a client, a session to a server is visible via RSessionBase, and to the server a session with a client is implemented in class CSession.

As generating messages in a client program can be tedious and error-prone it is common that an interface is implemented, which allows client programs to access a server via a DLL rather than the actual server (Figure 6.10). That eases the design task of the application developer, because an interface that is easier to use can be offered for the programmer. This of course can also be problematic because it is not straightforward to determine how much communication is caused by the operation.

Let us next consider server implementation assuming that a server was implemented that would manage the selection of the next question and answer in the sample Symbian application given in Chapter 3. The following code could be used, assuming that the client-side session was implemented in class RQASession:

```
EXPORT_C TInt RQASession::GetQuestionId()
    {
    const TAny * p[KMaxMessageArguments];
    return SendReceive(EQuestionRequestCode, p);
    }
EXPORT_C TInt RQASession::GetAnswerId()
    {
    const TAny * p[KMaxMessageArguments];
    return SendReceive(EAnswerRequestCode, p);
    }
```

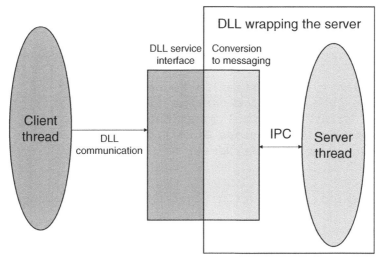

Figure 6.10   Wrapping server interface with a DLL

In every session, an operation called `ServiceL` is introduced, which acts as the counterpart of `RunL` in the active object framework. In `ServiceL` of a session (`CQASession` in the sample program) running in the server (`CQAServer`), a switch-case statement would then be given that would further call the actual operations. Code also exemplifies the use of parameters. Internally, this parameter passing will be implemented in terms of the normal inter-thread communication mechanism. In this particular case, assuming that only the above operations would be used, the following code could be given:

```
void CQASession::ServiceL(const RMessage & aMessage)
    {
    switch (aMessage.Function())
        {
        case EQuestionRequestCode:
            GetQuestionId(aMessage);
            break;
        case EAnswerRequestCode:
            GetAnswerId(aMessage);
            break;
        default:
            _LIT(KPanic, "QAServer");
            aMessage.Panic(KPanic, KErrNotSupported);
        }
    }
```

The actual message handling is then implemented in methods `GetQuestionId` and `GetAnswerId`, which could be given in the following fashion. In the listing, class `TPckgC` is a thin, type-safe template class derived from `TPtrC` that is templated on the type to be packaged (Stitchbury 2004):

```
void CQASession::GetQuestionId(const RMessage & aMessage)
    {
    iQuestion = (++iQuestion) % KQuestionCount;
    // Wrap integer to the message format.
    TPckgC <TInt> valueDes(iQuestion);
    aMessage.WriteL(aMessage.Ptr0(), valueDes);
    aMessage.Complete(KErrNone);
    }

void CQASession::GetAnswerId(const RMessage & aMessage)
    {
    iAnswer = (++iAnswer) % KAnswerCount;
    TPckgC <TInt> valueDes(iAnswer);
    aMessage.WriteL(aMessage.Ptr0(), valueDes);
    aMessage.Complete(KErrNone);
    }
```

The result of this approach is that resources are visible via method calls, and from the programmer's perspective it appears similar to programming using a monolithic kernel (Figure 6.11). However, the cost of resource access can be more expensive, since additional context switches may take place.

Additional consideration can be given for establishing a connection to open a session to the server and for closing the connection (named as `CreateSession`, `Open`, `Connect`; `TerminateSession`, `Close` for example). For instance, it may

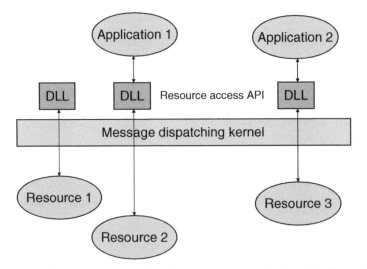

Figure 6.11   DLLs as an interface to resources in Symbian kernel

be necessary to introduce some logic to determine whether or not the server is already running. In addition, if a complex startup procedure is a necessity, where for instance some hardware resource is initiated, it is possible that version-specific variations as well as differences in emulators and devices imply difficulties. For instance, the sample server given by Stitchbury (2004) contains code that is different depending on the used environment due to the differences in threading models in the emulator and the actual device. Further complexity is added to the scheme when addressing the use of services, whose execution is preferably asynchronous. Then, support for asynchronous behaviors is also required of the client, implying that an active object is implemented to handle incoming responses, for instance.

Considering servers as event handlers, which active objects fundamentally are, leads to a message passing architecture where the kernel acts as the message dispatcher. Figure 6.12 illustrates the use of messages, and the role of the client of the server, the server itself, and the kernel that manages their connection (Stitchbury 2004).

6.4.3   Adapting Servers to Hardware

Similarly to other parts of Symbian OS needing adaptation, also hardware adaptation is managed with plugin components. As an example, let us consider perhaps the most important types of plugins related to communication hardware. In this domain, the different plugins are referred to as *communication subsystems* (CSY), *telephony*

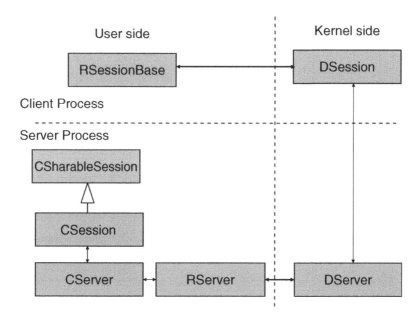

Figure 6.12   Role of kernel in Symbian client–server framework

subsystems (TSY), *protocol modules* (PRT), and *message type modules* (MTM), used for the following purposes.

1. *CSY modules* are used for specializing communications-related aspects to hardware and the used protocols. A CSY module communicates directly with a physical device driver. A CSY module can use other CSYs. For instance, the IrDA CSY module that implements the IrCOMM interface to infrared (IR) physical device driver also uses the serial device ECUART CSY module.
2. *TSY modules* are the type of plugins that are used for adapting to different types of telephony hardware.
3. *PRT modules* are the central modules used in protocol implementation. When a certain type of protocol is needed a server creates an instance of a suitable PRT module. Examples of PRT modules include `TCPIP.PRT` and `BT.PRT`.
4. *MTM modules* are used for handling different types of messages using the same framework.

Figure 6.13 illustrates how these modules are used in the implementation stacks of Bluetooth and TCP/IP with point-to-point protocol (PPP) (Simpson 1996) facilities. In Bluetooth, only one module (`BT.PRT`) is used, whereas TCP/IP with PPP requires using a number of modules of different type (`TCP.PRT`, `HAYES.TSY`, and `ECUART.CSY`). Further complexity is added when using wireless application protocol (WAP), where multiple PRT modules are used (`WAPPROT.PRT`, `SMSPROT.PRT`) (Jipping 2002).

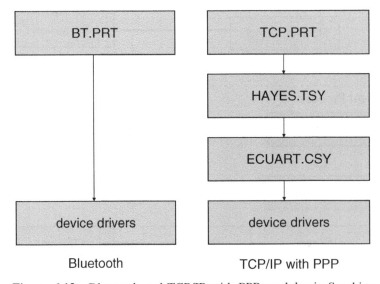

Figure 6.13   Bluetooth and TCP/IP with PPP modules in Symbian

Similar discussion can be given to multimedia codec adapters and cameras, as well as a number of other features where abstract functions are provided to applications, but low-level implementations vary. Furthermore, device drivers of Symbian OS are commonly implemented using logical and physical drivers, as already discussed. The former includes the common functions of the driver, and the latter is for hiding the details of the particular hardware implementation. A common design goal is to implement only a minimal physical driver, and keep it as thin as possible. In contrast, the logical driver then hosts the majority of functions that are assumed to be necessary in all different hardware configurations. For obvious reasons, this contributes to an increased compatibility between devices. Still, problems similar to those associated with Java can be encountered when porting Symbian applications from one device to another.

6.4.4 Problems with Servers

As already discussed, introducing a message passing architecture in a device that has restricted resources can be harmful to performance. In the case of Symbian servers, some optimizations have been introduced, where a number of servers run inside the same process to avoid superfluous copying of data in communication stacks for instance.

Another detail that can be problematic is the required startup sequence. As the behavior of some servers is intermixed, their startup order must be carefully designed. Moreover, when performing the boot, all servers are initializing themselves, and it is possible that they unnecessarily await acknowledgments from each other's successful startup, which in turn again results in delays. Optimally, a design is introduced where the startup sequence can be centrally controlled so that it can be changed with relative ease.

When a server runs into an internal problem, it may have a hard time recovering. The error may already have been propagated to other parts of the system and even if the server that first ran into the error could boot itself, others may still be using erroneous data. Failures in booting the server in turn can lead to unavailability of some of the hardware characteristics of the device, as in many cases each new device needs a controlling server.

Finally, one problem is that Symbian OS has been evolving rapidly. Therefore, changes in different parts of the system have sometimes led to restructuring of server-related features. This results in differences in the code base that can be used in them, but is next to unavoidable.

6.5 Summary

- Resource management can – and usually should – be encapsulated in specialized modules for eased design and maintenance. While this consumes resources, otherwise it is impossible to have any control over them.

- Several competing concerns to be taken into account when designing resource management:

  - Extensions and adaptations, which can be applied to low-level software, middleware, and executables,
  - Performance,
  - Energy consumption,
  - Internal resource management.

- Balance between offering possibilities for optimization and encapsulation is an essential design goal.
- Microkernel and monolithic operating system design are both possible. Moreover, they may in fact appear similar to the application developer, when resources are accessed via APIs in both forms in any case. However, their non-functional characteristics are different.

  - Microkernel introduces some overhead, but the parts of the system can be replaced at least relatively easily. The design of concurrent executions is eased, as the different executions reside in different components.
  - Monolithic design can become a burden to maintain in the long run, but the intimate connection of the different elements of the system usually results in improved performance. The design of concurrent executions can be difficult, as only relatively primitive facilities may be available. Moreover, debugging can also be difficult.

- As resource manager software is constantly active, internal resource handling in such systems should be carefully designed.
- Lowest-priority thread is commonly used for housekeeping and resource management activities.
- MIDP Java relies on the use of standard interfaces when accessing the resources. Actual implementation of the operations can be vendor specific, and the underlying operating system properties are hidden from the Java developer.
- Symbian OS relies on the use of message-passing architecture, where resources are encapsulated in servers.
- Porting to mobile devices is commonly made difficult due to different interfaces that are available to different resources. Moreover, implementations of standard interfaces sometimes vary. Furthermore, even if a certain hardware interface does exist (e.g. Bluetooth), the associated software interface may not be available for a programmer.

6.6 Exercises

1. As already discussed, using a message-passing architecture for resources often results in lack of performance. What kinds of benefits does it offer for the

developer in exchange, when compared to a monolithic design? Consider at least the following aspects:

(a) Memory consumption.
(b) Debugging.
(c) Maintenance.

2. What kind of a design is needed for implementing MIDP Java interfaces on top of Symbian server-based resource management strategy?

3. What would be required for implementing C/C++ standard interfaces on top of Symbian OS server managed resources?

4. What benefits would be offered by a system where all resources were included in the operating system kernel instead of being distributed to different resource managers?

5. Consider a license manager software system that contains a database of licenses and methods of payment for them. Each license-capable application can call a special method `CheckLicence(PaymentType)` which checks whether the license is available, and if not, the license manager purchases the license. What parts of the system should be implemented with plugin techniques? What would the different plugins implement?

6. In what situations would it make sense to offer several different interfaces to the same resource?

7. What types of features could be implemented with aspect-oriented techniques in mobile devices, assuming that the restrictions on memory usage were relaxed?

8. Consider using a DLL or a server that only runs when the application is active in a Symbian application. What differences exist? How would the differences be visible to the application developer? In what situations should the server solution be used?

9. Convert the sample Symbian model given in Chapter 4 to a server. What type of a server (always on, turned on only when needed, ...) would be a suitable implementation? What could be gained (or lost) if some other type of server implementation was used? How much is the execution time of the application affected?

7

Networking

7.1 Introduction

The fundamental purpose of a mobile device is to liberate users from restrictions of place. Therefore, it is only natural that the connection to a fixed network is gaining more and more interest in the domain of applications, and the role of a mobile device as an execution environment for local applications only is becoming less attractive. Furthermore, assuming that it is the contents and data that users are consuming, it is obvious that they should be loaded over the network in some form or another, as communication is the main purpose of many mobile devices.

An obvious consequence of the above is that one can propose the use of mobile devices as extensions of any networking system. For instance Pernici (2006) proposes a number of aspects that enable, allow, and support such extensions. Moreover, communications facilities associated with mobile techniques, discussed by Schiller (2003), for instance, have gained interest. In the scope of this chapter, however, we place the focus on the internals of a mobile device, and overlook the design of services that may be located in the network but can be specialized or extended for mobile devices. Therefore, while addressing the properties of networks and services, the viewpoint is that of an application developer for a mobile phone, not that of a service provider or service developer. Another consequence of the above is that in many ways, networking features appear as yet another resource that is capable of communicating with the surroundings. For the issues related to distribution, the reader is referred to Andersson et al. (2006), Coulouris et al. (2001), Mullender (1993), and Tanenbaum and van Steen (2002).

7.1.1 Basic Connectivity

When using mobile devices in a networking application, the limitations and practicalities commonly associated with the development of distributed systems and Internet programming remain valid. For instance, the basic properties, such as openness, bear similar importance in design, and the question is fundamentally to create

Programming Mobile Devices: An Introduction for Practitioners   Tommi Mikkonen
© 2007 John Wiley & Sons, Ltd

a system of nodes that are able to communicate with each other using some media, although in terms of resources, scalability, and at least partly transparency the situation changes. One can consider the possibility to use services of systems located on the network as yet another layer of facilities, like memory, as hinted in Chapter 1. While accessing data in this layer is more expensive in terms of time and actual payments, computing and storing resources can be vastly increased; servers in the network need not obey the restrictions applicable to software running in the mobile device, but they can be connected to constant electricity and run large and sophisticated software. As for transparency, it can be meaningful for the user to be able to tell the difference between local and remote features for cost awareness and offline working in an airplane, for instance.

An additional characteristic of the wireless domain, especially in connection with cellular networks, is an additional piece of networking equipment that acts as a bridge between the fixed and wireless networks, and uses bearers that are specialized to this kind of a setting. Such techniques have been used, for instance, in a wireless extension of Corba called Wireless Corba (Black et al. 2001) and in wireless application protocol (WAP) (WAP Forum, 2001), where a gateway implements the translation between the protocols used in the wireless domain and the fixed Internet. In such a context, however, only lower level protocols usually vary, not the protocols visible to the application. Therefore, no changes are implied to the applications, apart from measures taken for other reasons, such as to hide the slowness of the connection to the network.

7.1.2 Stateful and Stateless Systems

In a networking application, several architectural decisions can be taken. In many cases the main driver of the decision is to define the boundary between the parts of the system that run in the device and those that run in the server. While technical restrictions obviously impose constraints on the selected implementation technique, properties and functions of the application often force the selection of a certain implementation technology.

Two different types of systems can be built with networking facilities depending on whether or not stations of the system should remember previous events, or simply always provide the operation in a similar fashion, assuming that all operations are self-contained.

- *Stateful systems* preserve information related to their past operations and associated internal state. This allows the creation of series of related operations. Then, in many implementations there is a global state that is distributed to all the systems needed for running the application. In practice, one can often consider the existence of sessions as a state for an application.
- *Stateless systems* simply execute some routines generating output based on certain input, disregarding what they have previously executed. Then, when running

an application, there needs to be only one subsystem – or sometimes only the user – that maintains the state of the application, and uses services offered by stateless components.

With a state that is stored in the system, applications get more complex, because their behavior may (and usually does) depend on the internal state. Following the paradigm of conventional distributed systems where all participants are allowed to have an internal state, one part of a system is then commonly run in users' mobile devices acting as clients, and the rest of the system is run in a server that can reside in the fixed network. Then, the state of the application is constituted by the internal states of multiple computers, some of which can be mobile devices, although the state inside clients is usually minimal in comparison to the server. The role of the server can be to enable centralized functions and associated state that cannot be achieved when only the client is used.

Abstractions that are commonly used for implementing applications of this type are based on some middleware system, such as Java's Remote Method Interface (RMI) (Sun Microsystems, 1997). Lately, also Web Services (Singh et al. 2004) have gained interest in the scope of mobile systems, as the same services could then be used with both mobile devices and the fixed Internet, assuming that the services were designed to fit alternating screen sizes.

When restricting the discussion to cases where the state of the application can be kept local or partitioned such that requests and reply messages over a stateless protocol, such as HTTP, are a sufficient implementation mechanism, the functions of the system can often be simplified to browsing in the mobile setting. While seemingly impractical, the browser seems to have become an important technique for implementing distributed applications. The reasons are that browsers have been widely deployed also to mobile devices, and that they can be extended with plugins that are allowed to execute downloaded programs. Furthermore, applications have found their way to use browsers' facilities for implementing at least a degenerated form of statefulness.

One can consider two different alternatives for the execution environment. One is to always run all programs in the server and to download only the result to a mobile device, thus benefiting from improved resources of the server, and the other is to first download the application to the mobile device in some packed format, then run it, and only upload the result back to the server. The selection between these is probably application-dependent, as the need for communications with the user plays an important role. In some cases, it may even be practical to implement an application-specific browser that uses protocols optimized for the purposes of that particular application. For instance, the MUPE (Multi-User Publishing Environment) platform relies on the use of dynamic mobile Java functionality and class handling embedded in XML, as discussed by Suomela et al. (2004).

A downside of the browser-based approach is that due to long travel-through times for request–reply pairs, it can take a long time to download something from the Web. The problem is worsened if several requests are to be made before anything can be made visible to the user. An additional downside is that without the state stored in the application, all communication must contain information about the past actions if they are relevant for the behavior, which in turn increases network load. Other implementations include the use of hidden fields that are loaded but not displayed and cookies that are stored in client computers, but which may be restricted in their size.

For obvious reasons, networking applications in mobile devices need some state information. At the very least, they must be capable of learning whether or not network connectivity is available. However, due to reasons related to loss of connection in the wireless setting, for instance, incorporating too much state in a session can lead to difficulties, especially if the session cannot be recovered later on if the connection breaks. Another practical detail is that testing of stateful systems can become more difficult, as in many cases both the behavior in the correct state as well as in an illegal state should be addressed when validating the system.

7.1.3   Infrastructure Assisted or Ad-hoc Networking?

A particular feature affecting the architecture of a networking application is whether it is based on a certain fixed and managed infrastructure, or a formation of independent stations that form the network in an ad-hoc fashion. In principle, both can be implemented on top of the connectivity described above. However, the more one implements middleware in the form of wireless Corba, WAP, or mainstream browsing facilities, for instance, the more likely it is that the application is necessarily based on fixed infrastructure, whereas ad-hoc environment in contrast seems to favor lighter infrastructure in the mobile setting, where mobile devices communicate with each other using some low-range radio protocol, like Bluetooth, for instance.

With a fixed infrastructure, responsibility for certain actions can be allocated to a certain network unit. Similarly, when a device enters an infrastructure network, it often sends a message, which registers the device to the network. This eases issues like who can access what, and information regarding the location of certain services. From the application perspective, the application can often adapt an implementation where either the device initiates all operations, as is the case with browsing, or the device waits for the environment to perform some operation, for example sending a message to the device. Moreover, the devices can often manage changes in the environment themselves. For instance, it is common that devices themselves know how to communicate with the network even when the home network cannot be accessed, and the programmer seldom has to take this into account in the application. To summarize, a lot of the facilities are built into the device, in particular when regarding the infrastructure needed for communications, and the user can simply apply them.

In an ad-hoc network, responsibilities can be assumed on the fly as devices in an ad-hoc network enter and leave the network uncontrollably. For obvious reasons, locating a certain service, such as printing for instance, should be performed via some service discovery routine, similarly to all the other services one wishes to receive. Moreover, one may also wish to offer services to others, which allows one to create and provide custom services using the device. In general, one can consider the following characteristics (Kortuem et al. 2002):

1. Networks are self-organizing in the sense that the topology of the network changes constantly as nodes enter and leave the network, and move in it, thus creating new connections.
2. Networks are decentralized; all nodes are equally important.
3. Networks are also highly dynamic because nodes may move frequently and independently.

For devices that only benefit from services offered by other devices, the situation need not be different from using the infrastructure, apart from the fact that preinstalled support may exist only for preplanned services, and custom services may require their own custom client. However, if a device offers services to other devices, there obviously must be some support for serving other devices.

7.2   Design Patterns for Networking Environment

In the following, we propose some commonly applicable ideas in the design of a networking application in a mobile device. The goal is to address only the internals of the device, and we will overlook the consequences of the designs to the hosting infrastructure.

Use a networking wrapper. While programming environments commonly offer an extensive set of interfaces for accessing the network, they have not been optimized for any particular application. However, when composing a particular system, where the network plays a specific role, a lot of the complexities of the interfaces are not needed in the first place. For instance, defining a Bluetooth connection in general requires a number of details to be negotiated and set, whereas applications mostly use only a subset of these. Therefore, it is better to hide the rest of the details behind the wrapper. Moreover, the wrapper can also be used for other purposes, like enabling altering the used networking facilities. Using a wrapper will also ease porting of the application to some other environment, as it separates platform-specific parts from the application.

Consider treating networking features similarly to resources. As already discussed, the facilities of the network can be considered yet another resource, which is an extension to the use of wrapper discussed above. Therefore, at the level of software, it is sometimes natural to implement a corresponding resource manager that helps in working with the resource. For the parts that are more or less standard software infrastructure, the manager can be provided by the networking infrastructure of the mobile device, like Socket or Telephony Server in Symbian OS, but for more abstract or specialized resources, an additional abstraction can bring increased value.

Adding an interface to an application-dependent resource manager can make it appear similar to infrastructure resource managers, with the same implementation-level technical investment in the form of threads, for instance. This eases their use, and can in fact lead to the creation of a new standard component that hides some of the complexities of the standard interface that must offer all the options. For instance, a device may contain a master resource controller that can be used for canceling all the established communication connections when necessary. As applications should in any case be implemented such that the loss of a connection results in minimal harm, the fact that the user can actively cancel executions does not cause additional risks.

Allocate a responsibility of networking operations to some particular execution unit. A common case is that a networking system makes accesses to the network while the user is using the device. In order to ease the design, a separate unit of execution, usually a thread but also an active object in the Symbian environment can be used, should be allocated for serving networking operations. This contributes to advanced separation of concerns, as one unit of execution serves the user and the other the network. Managing the mutual exclusion is then a task for the programmer.

When implementing a server, keep it separate from client. Especially when implementing ad-hoc systems, it is common that mobile devices also host services. Then, it is often beneficial to separate the server from the client for several reasons. First, when using the application the local client does not get unfair advantage in games if not desired. Second, tracing bugs will be eased when either the server or the client is to be updated, if there is no need to study their merge as a separate case. Moreover, different techniques can be used in the implementation. For instance, the server may be implemented with Symbian C++ and clients using mobile Java. Finally, it will be easier to implement different clients that will use the same server if clients and servers are not intertwined. However, implementing a special interface for local operations can be a practical necessity to reduce networking load. Still, one can consider using only one interface but different plugins, one for remote and one for local data.

Telecommunications features are commonly more restricted to some particular purpose, whereas data communications enables more freedom. In many ways, telecommunications and data communications features reflect their origins. Telecommunications features, such as establishing voice calls and delivering simple messages using SMS and MMS messaging (Bodic, 2003; Mallick, 2003), can easily be accomplished, but it is not always straightforward to build new applications on top of them, as the service is already intended for a person. In contrast, in data communications the user can compose a program that reads incoming messages and responds accordingly relatively freely and in particular using whatever format seems to be most convenient. Therefore, the latter seems to be more natural for applications. However, telecommunications features can adopt a special role when for example a conversational communication channel to the help desk of an application is needed. For instance, for a company that wants to extend its information system to all the salespeople, the use of telephony-based messaging for communication seems to be a superfluously complex and expensive solution. Furthermore, the existing information system is probably built around Web technologies, and using the same infrastructure is only rational. Instead, for participating in different kinds of voting organized in cooperation with a cellular operator that require sending a premium-paid SMS, it is obvious that messaging is a necessity due to the business model of the organizer. While principally a question of business models and applications, there are some technical restrictions, however. Firstly, establishing a communications channel for a computer-like connection requires some time, whereas an SMS can be sent in the background without user interventions. Secondly, assuming that the networking end will be active, it is difficult to get users' attention without a mechanism that would interrupt them. For instance, implementing an email system that immediately notifies the user about an incoming email via a mobile device either requires a messaging-based notification of an incoming email via telecommunications that triggers the download using computer-like data connection that is to be constructed separately for all emails, or keeping the connection alive and polling the status, which in turn consumes battery.

Consider push versus pull. Push services are a commonly used term for services that the network initiates in a mobile device. In contrast, pull services are those that the mobile device itself initiates. Many of the services that have been implemented in the Internet obey pull philosophy, and it has also become important in mobile devices, where for instance a browser can be used to pull information from the network. However, the traditional telecommunications services have been such that it is the network that takes the initiative, and informs the user that there now is an incoming call or text message. However, in many ways offering the network the possibility to inform about certain events is a necessity for more elaborate mobile applications.

Be prepared for loading times. As already discussed, mobile communications are often considerably slower than corresponding fixed communications. In many cases, this will be difficult to hide from the user. However, giving the feeling of being in control to the user should be ensured to create practical networking applications. Moreover, while technologies like Ajax, which builds on JavaScript, `XMLHttpRequest`, which is a way to implement an asynchronous server call in the Web environment, Cascading Style Sheets (CSS), and other Web technologies have introduced facilities for more responsive Web development, the effects have been implemented by downloading executable code like JavaScript from the network to the browser, which then appears more interactive due to the downloaded code. Therefore, they do not as such ease building more responsive applications, but may in fact lead to extended download times. At very least, the application must somehow indicate that the input has been detected, and that the request is being served. Still, it can be argued that Web-based applications that run inside a browser are in fact among the most versatile pieces of software that can be used in a mobile device, although they may not have been intended for such use by the original designer.

Consider proactiveness. In the Internet paradigm, communication usually takes place in real time. There is no need to wait for a convenient time to perform certain networking operations, except at the very lowest level when the limitations of implementation techniques become visible. However, when using mobile devices, the issue becomes more complicated. To begin with, there can be several options for networking, of which some may not be usable in some locations. Then, it may be possible to delay certain operations in a GPRS connected area until for instance faster 3G network or cheaper and faster WLAN becomes available, at least if the amount of data to be transmitted is large. Moreover, also other reasons can be served. For instance, it may be much more energy efficient to transmit large amounts of data when the radio connectivity is strong, and only the least possible amount of energy is consumed. Unfortunately, not all applications can benefit from the option to delay the execution of some operations. For instance, if the user is accessing some data from the network that is urgently needed, performing the action later is not an option. However, in some cases, for instance, delivery of news or broadcasts as well as certain synchronization operations can be performed without user invention as a background activity.

Adapt to the facilities of the device. Unfortunately, in many cases the options offered by a mobile device are somewhat limited. Therefore, for practical application development, a good starting point is to study the different design alternatives offered by the platform, and select one of them as the basis of the system.

7.3 Problems with Networking Facilities and Implementations

A technical problem in networking with mobile devices result from equipment needed for implementing the connectivity. Round-trip times are long in cases where the cellular network is used as a bearer. First, the device must interact with cellular equipment for equipment and user identification and access, then some other piece of equipment provides the access to the Internet, where the actual services usually reside. In connection with low bandwidth – O'Grady and O'Hare (2004) suggest the theoretical and practical bandwidths listed in Table 7.1 for GPRS and UMTS – and slow interaction, this results in relatively long response times. Additional challenges are introduced by the fact that the connection between a mobile device and a system installed in the network may require re-routing as the device moves in different networks.

Table 7.1 Throughput of GPRS and UMTS

	Theoretical	Practical
GPRS	115 kbps	30 kbps
UMTS	2 Mbps	300 kbps

Another problem is constituted by the number of available techniques, deriving from both data and telecommunications origins. There are at least four different ways to communicate using messages: fax for sending paper documents as messages, short message service (SMS), which offers a primitive messaging service originating from telecommunications, emailing system similar to that of the Internet, and multimedia messaging service (MMS), which is a technique composed of both telecommunications and data communications. There are two browsing stacks, one for WAP browsers and the other for Internet browsing using hypertext transfer protocol (HTTP). Furthermore, possible data communications media include data calls, GPRS, WCDMA, and wireless LAN, of which the latter can be used for similar purposes to Bluetooth and infrared communications, at least in some cases. Sorting out the right technologies for an application, as well as technologies that are available for a programmer in different types of devices, are a complicating factor. In practice, it is therefore often advisable to use bearer-independent data communications whenever applicable, and require justification only when deciding for some other choice. In fact, one can consider that a prime challenge related to networking abstractions in mobile devices is how to integrate (or unify) the different facilities into a comprehensive yet simple and usable component that could be used without considering the details of different techniques in applications.

A further known shortcoming of communications techniques is that one cannot implement a system where one device would be directly offering services to others using a mobile operator's network as the media, if the operator does not enable distribution of the device's IP address. However, even such applications can be implemented with an additional server residing in the network that takes care of addressing and routing.

Finally, the costs of the connection that is associated with using the operator network can be considered a problem. In particular, if no information regarding the cost of a service is available, it is difficult for a user to commit to using the service.

7.4   MIDP Java and Web Services

Web Services have become a common technology for networking applications. For MIDP Java, the way in which Web Services are to be used has been specified in JSR 172 (Ellis and Young, 2003). The specification only discusses the use of Web Services on the client end, which enables mobile devices to interact with services offered by the network, but not vice versa.

In this section, we give an overview on how Web Services are implemented in mobile Java. The goal is not to go into details of the implementation but to demonstrate the available facilities.

7.4.1   Web Services Overview

Web Services (e.g. Singh et al. 2004) form an infrastructure-based networking technique which has been gaining more and more foothold in the Internet (Figure 7.1).

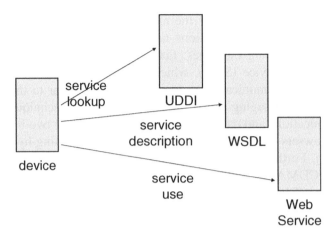

Figure 7.1 Using Web Services

Principally, the idea of the technique is to enable services that others can use in the composition of even more complex services relying in practice on TCP/IP and HTTP connectivity. While originally introduced for Internet-scale systems, Web Services have been introduced for the mobile domain as well (Pashtan, 2005). Being in use in many companies, the technology offers an option to compose services that can be accessed from mobile devices as well. Moreover, the cost of bringing the facilities to mobile devices is low, because the underlying implementation of the service can be the same. However, applicability is not always straightforward, as in many cases the goal is to perform machine-to-machine operations between systems of different infrastructure, not human-to-machine service implementation, which would most naturally be the case with mobile devices.

When designing a system relying on the use of Web Services using mobile phones, there is no particular application model that should be obeyed. In fact, as Web Services offer independent services to the application, technical issues like remote object references that must be considered when using a distributed object model become irrelevant, which simplifies the scheme. However, when using a service, one party adopts the role of a client (service user) and the other the role of a server (service provider). In practice, it is often a necessity to introduce some concurrency in the design of the application to keep the application responsive while Web Services are accessed over the network.

Finding Web Services is based on UDDI (Universal Description, Discovery and Integration), which commonly defines a directory structure located in a registry computer containing information on registered Web Services. In addition, other protocols have been proposed for finding Web Services that also support the ad-hoc environment. For instance, WSDD (Web Service Dynamic Discovery, or so-called WS-Discovery) protocol has been introduced for discovering Web Services in a dynamic network environment (Microsoft Corporation, 2005).

Communication takes place using SOAP protocol, which enables communication at a relatively high level of abstraction. From the programming perspective, a Web Service acts as a service, not as an object (or a collection of objects) that could be instantiated, which should be taken into account in the design. However, this predominantly reflects requirements to the server side, not to the client, as the situation is similar in the fixed Internet. Web Service specification does not define any standard bearer for the technology. However, in practice, HTTP is most commonly used due to its capability to travel through firewalls in the Internet.

7.4.2 Using Web Services with Mobile Java

MIDP Java adaptation introduces tool support for implementing client applications in accordance to the scheme illustrated in Figure 7.2 (Ellis and Young, 2003). Elements of the figure are:

1. the application that runs in a mobile device,
2. a stub that is used for making calls to Web Services and that can be generated automatically,
3. Service Provider Interface (SPI) that is used to enable the use of platform-independent stub generators,
4. a local Web Service client, developed using the WSDL file of the service, which will communicate with the server hosting the actual Web Service.

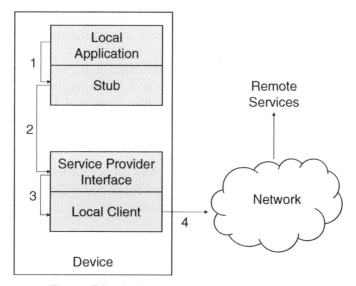

Figure 7.2   Stub and service provider interface

The operations that take place are the following:

1. A local application makes method calls to the stub.
2. The stub calls Web Services via the formally defined Service Provider Interface.
3. Service Provider Interface implementation uses the information received from the stub to open a connection to the remote Web Service.
4. The remote computer performs the requested operation, and once the operation is completed, the control returns to the local application.

Like most Java features, tool support for implementing Web Services in mobile Java has been based on APIs, which in this particular case consist of optional packages. The most important API packages and related tools are listed in the following.

- *Java API for XML processing* (JAXP) offers support for XML parsing (Suttor and Walsh, 2004). The parser is a non-validating parser that is intended to parse incoming XML documents and make the included data available. Furthermore, the parser is designed to parse an XML document as an input stream, rather than as a document tree.
- *Java API for XML-based RPC* (JAX-RPC) is an implementation of Remote Procedure Call (RPC) technology (Chinnici, 2002). The version targeted for mobile devices is a scaled-down version, and it is tailored to run in such a restricted environment.
- *Service Provider Interface* (SPI) is intended to allow the implementation of generated stubs in a compatible fashion.
- *Stub Generator*. JAX-RPC includes a stub generator that can be used for generating a client-side proxy that can be called by an application to place calls to a remote Web Service.
- *Java Architecture for XML Binding* (JAXB) offers a way to transfer an XML document to a set of Java classes and interfaces based on the document's XML schema (Fialli and Vajjhala, 2006). This enables the use of XML facilities using Java only.
- *SOAP with Attachments API for Java* (SAAJ) defines how to create and send SOAP messages (Jayanti and Hadley, 2001).
- *Java API for XML Registries* (JAXR) provides an access to standard registries, such as UDDI (Najmi, 2002). It contains operations for registering a service to a registry as well as for discovering services from one.

The actual development of a Web Service and its clients is usually assisted by supporting tools. For further discussion, the reader is referred to Maruyama et al. (2002) and Singh et al. (2004), for instance, where details of the approach have been discussed.

7.5   Symbian OS and Bluetooth Facilities

In connection with Symbian OS, we will study the use of Bluetooth, a cable replacement radio protocol for short-range communications. The Symbian environment offers sophisticated support for using Bluetooth, where a mobile device can both offer and use services to and from other devices. Both cases will be addressed in the discussion. Again, towards the end of the section, we will introduce also other networking options of Symbian OS.

7.5.1   Bluetooth Introduction

Bluetooth is a bearer and low-level radio technology introduced for short-range wireless communication, which started as a Bluetooth Special Interest Group, but which is now adopted as IEEE Standard 802.15.1. Originally intended as a cable replacement, several higher-level applications of the protocol have been proposed on top of this simple infrastructure. Offering a wide range of devices that have already been deployed, Bluetooth is a candidate for applications benefiting from ad-hoc and proximity-based networking, such as mobile games or different services associated with certain physical location. Frequency hopping is used to enable several Bluetooth connections in the same spatial area.

Bluetooth services are divided into service classes, associated with a 128-bit Universally Unique Identifier (UUID). which are a mechanism for ensuring compatibility between different types of devices. As an example, Table 7.2 lists some commonly used service classes and associated UUID values. Each service class is associated with a profile, which defines a selection of protocols and procedures that the devices associated with the profile must implement. As an example, Generic Access Profile (GAP) is the basis for all other profiles and defines how to establish a baseband link between two devices, Generic Object Exchange Protocol (GOEP) is used to transfer objects between two devices, Service Discovery Application Profile (SDAP) describes how an application should use Bluetooth's service discovery protocol to discover services on a remote device, and Serial Port Profile defines how virtual serial ports are created and how two devices are connected. In addition, each service class defines a set of attribute definitions, which define a 16-bit identifier for an attribute that can be advertised by a Bluetooth device. Some example attributes

Table 7.2   Some Bluetooth service classes

Service class	UUID
Serial port	0x1101
Dialup networking	0x1103
OBEX file transfer	0x1106
Headset	0x1108
Cordless telephony	0x1109

Table 7.3   Some Bluetooth attributes

Attribute definition	Attribute ID
Service Class ID List	0x0001
Service ID	0x0003

and their identifiers are listed in Table 7.3. These attributes are gathered to service records that specify the parameters needed for addressing a particular service in a certain device. There are three different power classes; class 1 is applicable for ranges up to 10 cm and uses 1 mW power for transmissions, class 2 for ranges up to 10 m and 2.5 mW, and class 3 for ranges up to 100 m using 100 mW. Obviously, class 2 is the most commonly used in mobile devices. Let us next consider the particularities of implementing applications using Bluetooth as the communication technique.

Bluetooth applications need not obey any particular application model, but they are usually based on the client–server paradigm, where the user of the service acts as the client and the provider as the server because of the forced asymmetric discovery of services. Moreover, there are restrictions on how Bluetooth-connected networks, so-called piconets, can be formed. For instance, there is a maximum of eight (one master and seven slaves) stations involved in one piconet. In addition, piconets can merge into scatternets, where a number of piconets can be involved. This, however, does not seem to be too common a use for the technology.

Bluetooth service lookup is based on device inquiry and a service discovery protocol SDP. It is used to locate devices that have their Bluetooth connection enabled (inquiry) and the services they offer (discovery). Device identity is based on a 48-bit Bluetooth address that uniquely identifies the device. As for services, the device that offers services for other devices has a special database that is used for registering services for others to discover.

Once a Bluetooth service has been selected, it can be contacted as if a serial port was used using RFCOMM protocol. On top of the serial port, it is possible to implement an application-specific protocol, or use standard protocols defined in Bluetooth profiles, which effectively define standard protocols for certain use cases. In practice, many applications that rely on Bluetooth use TCP/IP or UDP/IP socket communication due to their relative simplicity, which is also the approach introduced in the context of composing applications for mobile devices by Jipping (2002), for instance. In addition, more complex protocols, such as OBEX, introduced for exchanging objects, have been introduced, and they can be used for implementing higher-level operations that are suited for the underlying model of interaction. Furthermore, also additional features, such as AT commands for using a modem, have been introduced on top of the core technology.

7.5.2 Bluetooth in Symbian

Using Bluetooth in the Symbian environment requires several components. Firstly, a service database is used into which services can be registered so that other devices can query for them. Second, service discovery protocol SDP is used to discover the services. Finally, when a service has been found, it can be used. We will address these issues in the following.

Service Database

Offering Bluetooth services is based on a service discovery database that contains information that the device provides to others. When a device wishes to offer services, corresponding service records are added to the database using database server API. Methods are included for establishing a connection and closing it, querying version information, and for enabling counting of established connections.

The actual service database is characterized by class `RSdpDatabase`, where methods called `CreateServiceRecordL` and `DeleteServiceRecordL` are for creation and removal of service records in the service database. In addition, two methods called `UpdateAttributeL` and `DeleteAttributeL` have been provided for manipulating individual service attributes in a record. The creation of a database entry is illustrated in Figure 7.3, where four `UpdateAttributeL` calls are used. They include a protocol descriptor list, which is a data structure used to define protocol-related information, and service id, service name, and service description.

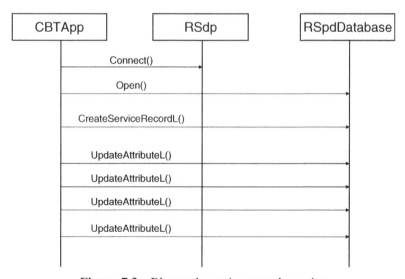

Figure 7.3 Bluetooth service record creation

Service Discovery

Bluetooth service discovery in Symbian OS is based on service discovery agent, `CSdpAgent`. When used, a service agent needs an object that is responsible for implementing interface `MSdpAgentNotifier` to be activated as the service discovery progresses. Callback method `NextRecordRequestComplete` is called when the latest request for a matching service search is complete. A handle received as a parameter can then be used for querying attributes. These results are then returned by `AttributeRequestResult`, and the completion status by `AttributeRequestComplete`. An agent can use a search pattern to find service identifiers. This is implemented with class `CSdpSearchPattern`, whose methods enable introduction or removal of service identifiers. With these facilities, a sample query for services would then be implemented as follows.

1. Create a notifier. This must be a class derived from `MSdpAgentNotifier`, and it must override the pure virtual methods discussed above.
2. Create an instance of class `CSdpAgent`.
3. Create a search pattern and add the service definitions that are to be discovered.
4. Install the search pattern in the agent with method `SetRecordFilterL`.
5. Signal the agent to start the search process by calling `NextRecordRequest` method.
6. Control will be dispatched to the Symbian Bluetooth framework that will activate associated callbacks.

Once service records have been found, attributes in them can be requested from the database using class `CSdpAttIdMatchList`, whose methods resemble those of service discovery (for instance `AddL`, `RemoveL`, and `Find`). The procedure to query for attributes is carried out as follows, assuming that the execution of the operation has proceeded to a point where notifier's method `NextRecordRequestComplete` signifies the service that was found.

1. Create a `CSdpAttrIdMatchList` object, which we will refer to as attribute pattern.
2. Add attribute identifiers to the attribute pattern.
3. Start attribute search using `AttributeRequest`.
4. Control is again dispatched to the Bluetooth framework. For every attribute found, `AttributeRequestResult` is called. When no more attributes that match the search pattern are found, callback method `AttributeRequestComplete` is called.

The execution of the service discovery process is illustrated in Figure 7.4, assuming that the right device has been selected and its address can be given as the parameter, and the associated search pattern has already been composed and

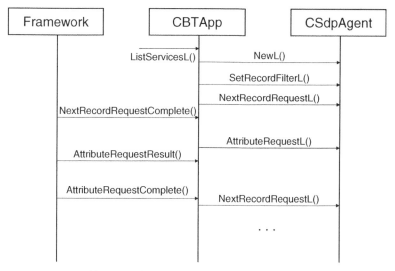

Figure 7.4 Bluetooth service discovery

it is passed as a parameter in method `SetRecordFilterL`. As shown, callback methods `NextRecordRequestComplete`, `AttributeRequestResult`, and
`AttributeRequestComplete` are used to activate application code when communication has been completed.

Payload Communications

As there are several protocols that can be run on top of a Bluetooth connection,
several procedures can be used. As an example, we discuss a primitive form of
using a Bluetooth service, sockets. When composing such a program, one must
connect to a Socket server and load the Bluetooth drivers. Following the guidelines
of Jipping (2002), for instance, this can be implemented as follows:

```
result = StartC32(); // Start comm server if needed.
if (result != KErrNone && result != KErrAlreadyExists)
    User::Leave(result);

result = socksvr.Connect(); // Connect to socket server.
if (result != KErrNone) User::Leave(result);

// Load protocol.
result = socksvr.FindProtocol(_L("BtLinkManager"),
                              protocolInfo);
if (result != KErrNone) User::Leave(result);
```

Opening and configuring a Bluetooth socket is similar to that of other sockets. Clients can connect to sockets using `RSocket::Connect` method, which

requires a valid address and service specification to which to connect the socket. This can take several forms, depending on the format in which the address is given (TBTSockAddr, TInquirySockAddr, or protocol-specific address). Finally, receivers of Bluetooth sockets must first bind an opened socket to a local address, then set up a listening queue for that socket, and finally open a new socket to connect to an incoming connection. To receive a connection, the service must call the listening socket's Accept method with a new socket, When Accept returns, the socket will be defined, and the two endpoints will have a data channel over which they can communicate. Closing a socket is performed by stopping the I/O requests associated with the socket, and shutting down first the socket and then the server.

7.6 Summary

- Network resources form another layer of facilities and resources for applications.
- Longish travel-through times due to complex network infrastructure and properties of wireless connectivity when using cellular protocols.
- Due to the restrictions of computer-like communications and messaging, some applications require both types of facilities to be used for best user experience.
- A common implementation of a networking application requires a resource manager, which handles both events originating from the network as well as those requested by the device. This resource manager effectively acts as an event handler not unlike those used for resources located in the device. In practice, this can be a thread or an active object.
- A key design decision is whether an application is stateless or stateful.
  - Stateless applications can be easier to use and compose, but transactions can require additional data elements.
  - Stateful applications may require less data communications, but can be harder to use and compose.
- Sophisticated communication facilities are offered by both MIDP Java and Symbian OS. In practice, designers often select the solutions that are best suited for the larger-scale good, not those that are optimal for mobile devices only.

7.7 Exercises

1. How should messaging and computer-like communications be used for the definition of an email system? What choices should the user be allowed to make in order to save battery? What choices require data in the server and what in the client end of the system?
2. Consider a system where the user's home computer is connected to the Internet using a broadband connection, and the computer does the actual browsing. When the user browses the Internet with a phone, the communication is always directed

to the home computer, which then interacts with other computers. When the home computer receives a reply to its query, it converts the resulting Web page to a JPEG and a collection of links. How much bandwidth can be saved in wireless communications using this approach?

3. Sketch an application that benefits from the concept of a location. What kinds of requirements does the application have on accuracy of the information? How about the performance, i.e., how fast should location data be available? How much data should be associated with a given location?

4. Study Symbian OS Bluetooth facilities. What kind of an API could be introduced to simplify the use of Bluetooth for client–server applications?

5. How should active objects be used in a Symbian implementation of a networking application to avoid two-thread pattern?

6. How should a networking application be designed in order to enable a system where one device would act as a host and others would use its services?

7. How would you compose a system where long-lasting operations, like sending an MMS, do not block user activity? Compare possible MIDP Java and Symbian OS implementations.

8

Security

8.1 Overview

In the PC and Internet environment, security-related issues are gaining more and more attention. Viruses and spam email have become a constantly worsening problem requiring user attention in the form of continuous loading of upgrades and updating spam filters, for instance. In addition to losing information, one can also lose money due to rerouted modem connections, for instance. As a solution, one must install firewalls and virus protection software to protect the computer.

For mobile devices, the cost of using a connection is well beyond that of a fixed Internet connection, at least potentially. Moreover, network operators and device manufacturers play a stronger role in the mobile setting than has been commonly assumed with the PC environment and fixed Internet connections. As a consequence, this has led to more elaborate mechanisms for security, where applications can have a wide range of rights to access system resources and privileges of the user can be restricted.

Fundamentally, the purpose of security features is to prevent unauthorized access to data and the features of the device and the introduction of superfluous costs or downgrading the resources. For resources that are accessed using network connections, a trustworthy implementation should be secure. This can be handled with the same encryption protocols that are used in fixed networking, and we will omit such issues for brevity.

For the resources that are located in the device, access means that interfaces are provided for addressing the data. The definition of resources, as well as the ways to stop unauthorized access, then becomes the important issue, as spyware and viruses can mask themselves as useful programs. Because applications can have different levels of trust depending for instance on their origin – manufacturer installed applications can be considered more trustworthy than those installed over Bluetooth by an unknown party – the traditional approach, where user identity is used for determining whether or not some right is granted, is not adequate. Instead, it makes more sense to use other means based on rights granted to applications.

   Software security can be considered using different scopes. One scope is construction time, where the design of the system occurs in the first place. In this case developers will be responsible for making secure, or security-enabling, design decisions. Another scope is the development and deployment of software for a system where a number of security measures have already been implemented. In the following, we first address generic design practices that aim at the development of secure software. Then, we will discuss the required security infrastructure and the types of operations that will be performed. Towards the end of the chapter we will discuss security infrastructure in terms of MIDP Java and Symbian OS.

8.2   Secure Coding and Design

Secure coding and design has been gaining a lot of interest over recent years. Motivated by security problems of existing systems, groundwork for composing secure systems has been introduced, including in particular the work of Graff and van Wyk (2003) and Yoder and Barcalow (1997).

8.2.1   Mindset for Secure Design

Graff and van Wyk (2003) propose the following process for designing a secure software system:

1. *Assess the risks and threats.* What are the bad things that might happen, and what are the legal and regulatory requirements?
2. *Adopt a risk mitigation strategy.* Plan in advance how to manage the risks identified above.
3. *Construct a mental model to support the development.* For instance, one can consider analogies such as sandbox, jail, safe, or honeypot to facilitate development.
4. *Settle high-level technical issues (for instance stateless vs. stateful, use of privileges, etc.).* The definition of the behavior of the system at an abstract level.
5. *Select suitable security techniques to satisfy the requirements.* What particular techniques and technologies are to be used.
6. *Resolve operational issues.* Such issues may include, for instance, synchronization or application-level communication encryption.

Furthermore, Graff and van Wyk (2003) propose that security should be engineered in the system starting at the beginning of the development. The rationale is that later on it is difficult to introduce security measures that would cover the whole system to the fullest extent. In addition, one should design for simplicity even when attacking a complex problem of security. The more complexity one puts into a design, the harder it is to validate its appropriateness, not to mention the likelihood of programming and design flaws.

As with any design, it is not only the technical artefact that counts but also its validation. Similarly to testing that can be characterized as showing that the program works, which can be considered to produce relatively few errors, or as the process of finding the hidden problems in a program, which focuses on finding the bugs (Myers et al. 2004), the validation of security properties also requires a special attitude. The assumption is that there are security holes in the system, and it is up to the validation and verification personnel to find them. Finding bugs and problems in the development should be a positive event (Graff and van Wyk 2003; Whittaker and Thompson 2003).

8.2.2   Sample Security-Related Design Patterns

As security can be considered a quality requirement, having no single responsible component, its implementation often requires architecture- and design-level measures. In the following, we introduce some design patterns that according to Graff and van Wyk (2003), Whittaker and Thompson (2003), and Yoder and Barcalow (1997) can be used as elements of security.

Role-based Security. The purpose of role-based security is based on using roles as the basis for privileges. For instance, users can assume several roles, like superuser, user, guest, depending on their intentions. Moreover, it is possible to also associate privileges with other facilities than user. For instance, giving certain privileges to applications or installation packages, it is possible to rely on roles. In reality, most devices implement two different roles. One is unidentified role that basically allows one to make an emergency call. The other is the normal user privileges, which grant access to all the resources of the phone and for instance in Finland require a PIN code from the user. Moreover, it is possible to require additional identification when performing some operations, such as resetting the phone to factory settings. In the future, it is possible that additional roles will be introduced. For instance, being able to define a role for device management that would allow centralized management of all additional software of the phone could lead to more elaborate use of mobile devices in corporations. Currently, this is made difficult by the fact that the users can always override corporate information management's effort and installations. In addition, platforms can offer different capabilities to applications running in them, which makes applications role-enabled entities. We will return to this topic towards the end of the chapter.

Checkpoint. In analogy to military installations, a checkpoint is a particular location in a design that is used to check whether or not an access or an operation is allowed or should be interrupted. In software design, this can be interpreted to mean that all data related to authorization and identity checking, for instance, is encapsulated in one object. Obviously, there can be different types of checkpoints for different purposes, following the plugin principles for specialization. For instance, one type of checkpoint can be used for authenticating a user, and another type of checkpoint may be necessary for loading a dynamic library. A third use would be the check where the authority for the installation of an application is granted. However, there should be only one way to accomplish the operations, i.e., in all cases where user authentication is needed, the same mechanism is used, all loading of dynamic libraries is performed using the same procedure, and there is no other way to install an application but the application installer must always be used. For instance, MIDP Java's restrictions on using user-defined class loaders discussed earlier can be considered as an instance of this pattern. One more issue related to using checkpoints is whether separate programs are used for validation, meaning that validation is specific to some particular part of the system, or if validation is based on declarative issues associated with all program, in which case it is easier to modularize and change.

Layered Security for Communication. The purpose of layered security for communication is to guide the designer to implement layers of security for all layers of communication instead of addressing all security in a single security model. In many cases, it is enough to implement an application-specific security layer, because existing components exist for lower-level protocols. For instance, when implementing a networking application, it is likely that the underlying networking subsystem and database introduce some security mechanism of their own. However, if no such libraries are available, one should be encouraged to build security systems for the different layers separately, as the design and verification of all the layers separately is usually easier than creating a single module that is responsible for all security features. Moreover, also the general principles of separation of concerns are better served in this design, contributing to easier maintenance. Obviously, a high-level secure access layer in and out of the application is needed in order to allow it to exchange information with its environment. Still, before implementing this layer, one should consider whether or not application-level communication must be secured, or if encryption at lower-level communication already ensures privacy to the desired degree.

Limited View or Full View with Errors. A design issue not directly associated with security, but bearing some consequences to it, is how much of a system is revealed to its clients. In particular, are only the operations that are offered to a certain client at a particular time revealed, or are all the operations always visible? *Limited view* to a system is based on showing only operations that are enabled for a client in a particular case. For instance, the user interface of an application may only show actions that are currently available for this particular user, taking the state of the system into account. From the performance point of view, the application of this design solution can be problematic, because in principle all the data that is not the same in all situations for all users must be produced by automatic generation. However, no information on existing but currently inaccessible properties is revealed. In contrast to limited view, *full view with errors* is based on revealing all the possible operations that a system enables, disregarding whether or not they can or are authorized to be executed in the current situation. Then, it is a responsibility of the client to recover from the error, and act accordingly. While the performance of this approach is usually better than that of the limited view pattern, making also confidential operations visible enables a hostile client to attack the operations more directly than if their details are not known.

Wrapping. While commonly associated with the role as an adaptor, wrappers can also be used as a security mechanism, as pointed out by Graff and van Wyk (2003). Obviously, depending on the options offered by the underlying system, wrappers can become very complex entities that perform a number of operations on behalf of the actual system. When using wrapping, there are a number of security-related aspects that can be taken into account before performing the actual call to the original component. Sample additional measures include the following:

- Sanity checking of parameters to guard against buffer overflows and parsing error attacks.
- The option to create a more restricted run-time environment in order to host the execution in less capable mode.
- Starting the system for the first time, allowing the wrapper to perform some extra activities related to security and other matters.
- Logging information.
- Adding pre-execution or post-execution code to the system.
- Intercepting the startup of an application.

8.3   Infrastructure for Enabling Secured Execution

In addition to designers, also other stakeholders are associated with security features, including users, network operators, and content providers and distributors. To them, the goal is not to design systems in a secure manner, but to introduce software that uses the available facilities in accordance to in-built security concepts.

8.3.1   Goals for Security Features

In the following, we discuss the different goals that are to be achieved with security features. The goals we will be looking at include confidentiality, integrity, authentication, authorization, and non-repudiation.

Confidentiality. Confidentiality is about what can be read. Confidentiality can be achieved by cryptographically transforming the original data into a form that cannot be read before retransformation back to the original form. Both operations are realized with a parameterized transformation, where the parameter, referred to as the key, is kept a secret. Transformations are commonly referred to as encryption and decryption. Depending if the key is the same or different, the terms symmetric and asymmetric are used. The biggest benefit of the latter is that one key can be made public. Then, one can preserve the ability to generate protected content with private key, and allow others to decrypt it with public key. Confidentiality becomes an issue at least in the following use cases:

- Download and installation. Downloaded applications may be something that are not to be revealed to others. Furthermore, further distribution of downloaded applications may be something that must be disabled.
- Communication. For obvious reasons, any communication that is performed should remain confidential, or at very least, there should be a mechanism to enable this when needed.
- User and application data. Confidential information in the device should not be exposed. For example, certain files can be hidden from the rest of the system, so that they cannot be accessed by other applications. Implementing this requires support from the underlying operating system. However, if no such support is provided, the application can internally use encryption to achieve confidentiality.

Integrity. The goal of integrity is to keep information intact. This can be achieved with cryptographic transforms and an associated key, together with some extra text to verify the integrity of the original text. Often, symmetric encryption and decryption, or simple checksums, depending on the case, are used. Integrity may become important in the following use cases in mobile devices:

- Download, installation, and startup. When a new application is loaded to a device, it seems rational to be able to check its integrity. Similarly, when an application is launched, its integrity can be checked to prevent tampering. For instance, checksums can be used as means of implementation.
- Communication. In order to ensure that communication has been completed in certain cases that bear transactional nature, integrity of communication packets can be validated.
- Run-time executables. When loading a piece of code from mass media to the device's memory for execution, its integrity can be validated to protect against tempering. Again, checksums and digital signing can be used.
- User and application data. Data integrity should also be protected in the device.

Authentication. Authentication is about an actor (the claimant) convincing another actor (the verifier) of its identity. Digital signatures based on asymmetric transformations are commonly used for this purpose. Authentication can be considered at least in the following cases:

- Access to resources. It may be a necessity to authenticate before the resources of a phone can be used.
- Download and installation. Incorporation of new software can require authentication to ensure that no software from an unknown source is installed.

Authorization. While often mixed with authentication, authorization is not about convincing the identity but getting permission to act. In many practical cases, authorization is associated to authentication, as with authentication it is possible to identify who gets the authority to perform some operations. As a result, implementation techniques are similar, and the following cases can be listed:

- Access to resources. After authentication, authority to use some resources can be granted. In many cases, this case is closely related to the right to use the device. For instance, authentication of SIM is commonly required before a user is authorized to access the operator's radio network. Also other facilities of the device can be protected in a similar fashion.
- Download and installation. Again, upon authentication, the authority is granted.

Non-repudiation. The purpose of non-repudiation is to ensure that it is impossible to deny certain actions afterwards, commonly using a combination of implementation techniques discussed above. Non-repudiation can be considered in the following cases:

- Download and installation. For downloads that result in billing, sufficient evidence of non-repudiation must be provided. However, these features are not implemented in the mobile device but in the server that enables the download. Non-repudiation may also be a necessity to perform for reasons related to warranty of the device; if one installs an application that has not been signed by a trustworthy third-party developers, changes in the terms of the warranty can be implied. Similarly, license managing software probably needs some kind of facilities for non-repudiation.
- Access to resources. Any action that can cause costs to the customer is at least potentially something that should be non-repudiatable.

For further discussion, the reader is referred to Gehrmann and Stahl (2006) and McGovern et al. (2006). Next, we introduce some practical facilities to achieve the above.

8.3.2 *Supporting Hardware and Software Facilities*

While in principle, one could assume an all-software implementation for security mechanisms, hardware can, and often should, play an important role. The reason is that if all the keys used in the cryptographic operations are stored in software, they could be tampered with by a malicious software. Moreover, when taking into account that security features are almost always an overhead from the user perspective – the user is more interested in getting an application running than for instance in associated digital rights management – using hardware acceleration can be considered a practical option.

As an example, Gehrmann and Stahl (2006) introduce an architecture based on cooperating hardware and software components for security purposes, which is targeted to mobile devices. The included security functions are listed in the following.

- Secure boot and software integrity check. The function is based on using a checking method stored into one-time programmable memory, which performs an integrity check when data and programs are loaded from flash. For more complex operators, also full installation integrity check is possible. Figure 8.1 illustrates how this can be performed.
- Secure control of debug and trace capabilities.
- Digital rights management. When new applications are installed in the device, their origins are authenticated, assuming that they will receive access to confidential parts of the device.
- IMEI protection and SIM lock, as one would expect. Maintaining this information in application software could more easily lead to malicious software accessing the data in an unauthorized fashion.
- Hardware cryptographic accelerators.
- Hardware-based random number generator.
- Cryptographic algorithm service.
- Public key infrastructure (PKI) support.
- Secure communication protocols, including GSM/GPRS/WCDMA security, TLS/SSL, IPsec, and Bluetooth/WLAN.

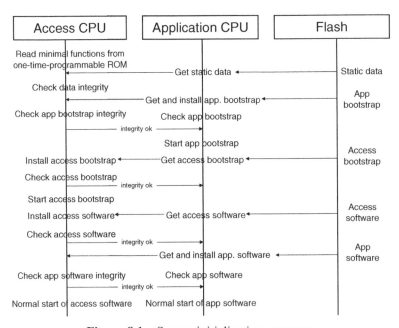

Figure 8.1   Secure initialization sequence

Gehrmann and Stahl (2006) also give an extended discussion on each of the functions, which explains the goals behind including this set of functions in the platform. Furthermore, the importance of trusted applications is highlighted.

8.4 Security Features in MIDP Java

Security features in MIDP Java have been divided to three categories. These categories, referred to as:

- low-level security,
- application-level security, and
- end-to-end security

are discussed in more detail in the following in dedicated subsections. Towards the end of the section, we also address problems associated with this design.

8.4.1 Low-Level Security

The main purpose of low-level security in Java is to protect the virtual machine by ensuring that no execution will interfere with the virtual machine. In common Java, this corresponds to class file verification, which is a computationally heavy operation to perform.

In CLDC-based Java where resources are scarce, this scheme has been simplified by including additional information, which assists in the verification process, in the generated Java archive by the development workstation. As a result, a new preverification phase has been added to the Java development workflow (Figure 8.2) in order to ease the processing of an application when the application is loaded.

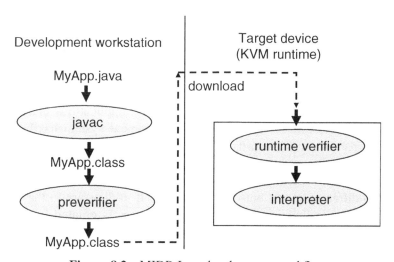

Figure 8.2 MIDP Java development workflow

When the virtual machine starts to load an application, the machine verifies the application using included additional data as assistance, and in case of problems abandons it.

8.4.2 Application-Level Security

Application-level security features of MIDP Java are based on the so-called sandbox security model. The model provides a good analogy for considering the security strategy. The goal is to keep the introduced software in sandboxes of their own, so that they will not interfere with each other's behavior.

In terms of an implementation, the sandbox approach of MIDP Java implies the following properties:

- Class files include standard Java applications that have been verified.
- A well-defined set of interfaces is offered to the developer, and the developer cannot introduce additional interfaces. Furthermore, extending these interfaces is forbidden.
- New application loading and management is handled by the virtual machine. Therefore, user-defined class loaders are forbidden, and there is only one, predefined way to load applications.
- Hardware and host platform features can be accessed only via the virtual machine and predefined interfaces. Changing these interfaces is forbidden.
- By default, applications can only load files from their own JAR file, and access resources from their own midlet package.

In practice, the above requires that all applications are run in separate virtual machines, or that the virtual machine implementation is such that it can handle the execution of different applications in a fashion where applications are granted to be isolated from one another. The latter results in a more complex implementation, and therefore many mobile devices only allow one Java application to be run at a time to avoid the instantiation of multiple virtual machines.

As the practical unit of security, MIDP Java relies on midlet suites. Applications are allowed to share resources with other applications that reside in the same suite, and user-defined libraries are enabled only for those applications that are located in the same suite. Moreover, also a record management system has been defined such that a separate set of records is offered to applications that are located in different midlet suites, whereas midlets in the same suite share files. As a result, a file named similarly refers to different files when the applications referring to it are located in different suites, but to the same file if midlets are in the same suite. Moreover, only controlled access to external resources is granted using operation getResourceAsStream. The situation is illustrated in Figure 8.3.

As already discussed, MIDP 2.0 introduces improved facilities for accessing devices' resources. The standard can be seen to define four levels of overlapping

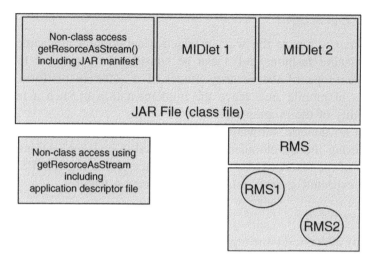

Figure 8.3   Midlet suite as a sandbox

sandboxes, and each of these levels can offer different privileges. These levels are the following:

1. *Device manufacturer* is usually granted all privileges to use the system, although there can be some operator-specific properties that should not be overridden by the manufacturer.
2. *Operator* role is similar to the device manufacturer, but this time manufacturer-specific properties should not be accessible.
3. *Trusted third party* can be granted access to the parts of the system that should not remain private for the device manufacturer or the operator.
4. *Untrusted party* only gets limited access to the facilities of the device. An implementation can be given where the user is asked for permission to perform an operation that is not directly authorized for the application, but which the application needs.

The author of a piece of software can be verified to belong to one of the above groups using certificates. Therefore, it is possible for the user to ensure the origins of the application. Moreover, when an application wishes to use resources it is not authorized to use, the device is allowed to ask for the authorization of the operation from the user. This allows the development of versatile applications even if all the certificates were not available.

As already discussed, no Java native interface is provided. Therefore, there is no way to circumvent the sandbox from Java applications. However, assuming that some software can be added to the native side of the hosting device, some 'opening' of the sandbox can be implemented.

8.4.3   End-to-End Security

End-to-end security means the way in which a networking application is implemented in a secure fashion, and it can be considered as an issue that does not fall within the scope of the security model of a software platform. Therefore, early versions of mobile Java leave the implementation of such a feature under the responsibility of the programmer.

Still, while no generic software architecture is provided, several facilities are offered for easing the development of secure end-to-end applications. Therefore, later versions of the standard also include interfaces for using HTTPS and SSL as well as other expected security-related communication features.

8.4.4   Problems

One commonly addressed issue regarding mobile Java and its security mechanism is that a number of important parts of the device are beyond the reach of applications. At first, this does not seem to be too essential a restriction, but as pointed out by Spolsky (2004), it is usually the data that is essential in a mobile device, and if no access is provided to it, developing of interesting personal applications is difficult. Although the situation has been changing due to the introduction of more liberal security schemes, it can be complicated to compose programs that address user data and are still portable across different devices. Moreover, although the user can often confirm an access also to restricted interfaces, constant authorization of execution can become frustrating.

Testing can also be considered problematic. In order to run as the final application, the system under test should behave similarly to it in all ways. However, assuming that the final application must be cryptographically signed to enable the verification and validation of its origins, it is only natural that at least some of the tests require a certified version of the application. This can lead to additional costs to acquire the certificate for the application as well as be slow if some other party performs the generation of the certificate. Furthermore, also minor bug fixes lead to new certificate generation, with similar downsides. Also, subcontracting for a main contractor that will have a fundamentally more permissive role can be complicated until the software nears completion.

A further problem is related to the control of programs and their distribution. For instance, assuming that an operator provides a mobile device, it is possible that also some restrictions on what software can be used on the device are introduced. This in turn reduces possibilities of implementing software that can be deployed in all phones.

8.5   Symbian OS Security Features

The way in which security has been incorporated in the Symbian platform has evolved over the years. The current set of security features includes run-time security in the form of capability-based platform security, checks that are performed at installation time, and facilities for performing secure communications. For clarity, we use the same grouping as with Java to discuss them, although no such grouping is explicitly mentioned in connection with Symbian OS.

8.5.1 Low-Level Security

Low-level security features of Symbian OS are not unlike those in many other computing systems. The underlying infrastructure protects different processes from each other using memory protection offered by the underlying hardware. Furthermore, the kernel is also protected from user programs, and they can only access the kernel's resources via a well-defined interface. In addition, some limits on memory consumption on stack and heap as well as on thread usage have been set to protect the system against malicious applications that may attack the system by reserving resources they are not planning to release.

Some run-time security features have also been defined, which are hard to categorize either as application-level or as low-level security features. However, they are related to the concept of platform security, which has been introduced as an application-level security mechanism, which we will address next.

8.5.2 Application-Level Security

A capability-based security scheme called platform security is a recently introduced Symbian OS facility. It has been introduced in order to allow fine-grained access to resources at run-time (Heath 2006; Shackman 2005a). In this scheme, the system is decomposed into security layers that have different capabilities, which give them a privilege to perform certain actions.

While the security layers of Symbian are not solely related to application-level security only but are used as a means for end-to-end security, the layers form an integral concept. Therefore, we treat then as one entity and do not distribute their effect to different levels of abstraction. The layers are characterized in the following:

- *High-level applications* are built on top of provided secure facilities.
- *Trusted environment (TE)* enables the use of communications, user interface, database, and security libraries in a reliable fashion.
- *Trusted computing base (TCB)* enables safe saving of data as well as facilities for identifying and authorizing software. In addition, this level also includes the hardware environment, although it is discussed separately. The different components of TCB automatically trust each other.
- *Trusted hardware base (THB)* enables safe initialization and execution of applications as well as secured hardware services.

In terms of an implementation, the ability to use the services of some other software artifacts is based on capabilities. It is the capabilities that form the basis for checking that permission to use services is given, and on top of which the above layering is implemented. Capabilities are given such that all the processes get minimum capabilities for implementing the tasks they are responsible for. In other words, the capability model can be interpreted such that capabilities define the extent to which a process is authorized to use the features of the system.

Several capabilities have been introduced. Some of the most important ones, together with identification on whether they are intended for system- or user-level functions, have been listed in Table 8.1. In addition to these technical artifacts, a number of processes have been defined for enabling security. These include authentication and certification processes and device management processes, to name a few.

The design principle that Symbian security features follow is that all processes have their own security level. In this scheme, processes that have high security requirements are usually those that are vital for the operating system, and less restricted security requirements are set to those parts of the system that can be added to the system by external parties later on. In addition, a small number of intermediate levels have been defined.

The implementation of the capability model is based on checking the capabilities when loading a dynamically linked library to a process or when contacting another process. For instance, consider the following cases. If a process with capability A dynamically loads a library that has capabilities A and B, the library code will be executed with capability A. However, if the library only contains capability B,

Table 8.1   Some common capabilities

System capabilities	Manufacturer set	`TCB`
		`AllFiles`
		`CommDD`

System capabilities	Extended set	`PowerMgmt`
		`MultimediaDD`
		`ReadDeviceData`
		`WriteDeviceData`
		`ProtServ`
		`Network Control`
		`DRM`
		`SurroundingsDD`

User capabilities	Basic set	`LocalServices`
		`UserEnvironment`
		`ReadUserData`
		`WriteUserData`
		`Location`
		`Network Services`

User capabilities	Other	Non-classified APIs

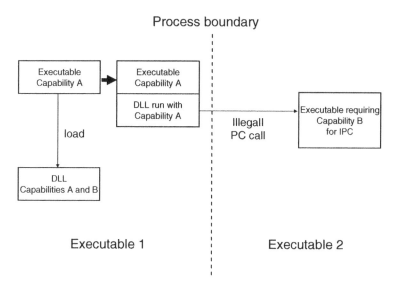

Figure 8.4 Inter-process communication and capabilities

loading will fail. Similarly, process communication can only be enabled if the capabilities match, which can sometimes lead to confusing errors. For instance, consider the case illustrated in Figure 8.4. Assuming that a DLL can be used in processes having capabilities A and B, but requires services from a process that has only capability B, executions where the DLL is loaded by processes having only capability A lead to failure. Because processes may require the use of certain capabilities, even if a DLL would in principle be able to communicate with some other process, inter-process communication may fail if the original process does not have the right capabilities. The effect of capabilities can be managed with general security settings when building ROM images. For instance certain (or all) capabilities can be automatically granted, or failed capability checks can be reported.

In addition to run-time security based on capabilities, some data-security-related features have been implemented. In particular, disk locations can be made private. As a result, directories can be made private to applications, which can then store their private data in them. Furthermore, application binaries are not visible to other applications. At the level of implementation, this has been implemented in terms of capability-based security. Capability to access an application's private data is granted to a minimal number of features. For instance the file server can manage applications' file use, and kernel, installer, and file server can access all executables.

8.5.3 End-to-End Security

When installing applications to a Symbian device, checks are performed that authenticate the application. However, even if the application cannot be authenticated, the user has been able to authorize the installation. However, since the introduction of capability-based security, only a limited authorization will be enabled. In other words, there are cases where composing a program for personal use becomes impossible without access to a suitable authorizing signing program, such as Symbian Signed.

Communication time security in Symbian can be based on the layered security scheme, although this is by no means enforced. The platform offers a number of secure protocols from which a suitable set can be picked. However, it is also possible to introduce application-specific crypting if necessary. Using this alternative is up to the designer.

8.5.4 Problems with Symbian OS Security Features

The most obvious problem associated with Symbian OS security features is the same as with mobile Java: restricted access to the system's resources can be demotivating for application developers. Further concerns can be raised on the future of application distribution and testing of applications at development time, which may require certification before an application can be tested.

A technical challenge associated with security features in Symbian is more complex testing. Because all interfaces are extended with capability definitions, integration testing gets a new dimension where the compatibility of different configurations of dynamically linked libraries in different executables is addressed. Moreover, measuring the coverage of such testing is hard, since if only code coverage is measured, it is possible that some combinations of different capabilities are overlooked.

8.6 Summary

- Security takes place at several levels of abstraction in mobile devices, including at least network communications, device's resources and data, and internally run programs.
- Install and run-time security features are commonly introduced to protect the device. Certificates can be used for verifying the author and the origins of a piece of software.
- Generic security patterns have been introduced, including for instance role-based security, checkpoint, layered security for communications, and wrapping.
- Several security schemes have been implemented in practice. As examples, we addressed two implementations.

  - MIDP Java application-level security is based on the sandbox model.
  - Symbian OS platform security relies on the use of capabilities that are related to provided privileges.

  Both schemes can be considered beneficial, as in principle they allow controlling the distribution of applications. On the downside, it is also possible to misuse this capability.

8.7   Exercises

1. Consider a license manager software that manages application licenses and transactions taken for paying the costs associated with the download and installation of applications. What kind of an architecture would be appropriate for the system? What parts of the system should be implemented as plugins? What parts of the system should be ciphered in the disk? What kinds of privileges would the software require in Java and in Symbian OS environments?

2. The current MIDP Java virtual machine runs only one midlet at a time. As a result, only one midlet at a time is being run in practice due to memory-related restrictions. What new security features would be needed in order to allow one virtual machine to run several midlets in parallel? Where would they be included in a virtual machine implementation?

3. What problems would arise if all software inside a mobile device was implemented with MIPD Java? What types of applications cannot be implemented due to the use of sandbox security?

4. Many standards require that it must always be possible to establish an emergency call. What kinds of designs would be possible for accomplishing this, assuming that viruses could downgrade the functions of the device?

5. Consider MIDP Java's and Symbian OS's run-time security mechanisms. What principal differences exist from the developer's viewpoint? How do these differences restrict application development?

6. How do the design of a DLL and a server differ in the Symbian environment when considering capability-based security?

7. What kinds of features could be implemented using the capability-based security for device management in the Symbian environment? What kinds of extensions would this require from the implementation?

References

Andersson E, Greenspun P and Grumet A 2006 *Software Engineering for Internet Applications*. MIT Press.

Babin S 2006 *Developing Software for Symbian OS: An Introduction to Creating Smartphone Applications in C++*. John Wiley & Sons, Ltd.

Barr M 1999 *Programming Embedded Systems in C and C++*. O'Reilly.

B'Far R 2005 *Mobile Computing Principles*. Cambridge University Press.

Black K, Currey J, Kangasharju J, Länsiö J and Raatikainen K 2001 Wireless access and terminal mobility in CORBA White paper, Highlander Engineering, Nokia, University of Helsinki.

Bloch J 2001 *Effective Java – Programming Language Guide*. Addison Wesley.

Bodic GL 2003 *Multimedia Messaging Service: An Engineering Approach to MMS*. John Wiley & Sons, Ltd.

Bosch J 2000 *Design and Use of Software Architecture. Adopting and Evolving a Product Line Approach*. Addison-Wesley.

Brown WJ, Malveau RC, McCormick HW and Mowbray TJ 1998 *AntiPatterns: Refactoring Software, Architectures, and Projects in Crisis*. John Wiley & Sons, Inc.

Buschmann F, Meunier R, Rohnert H, Sommerlad P and Stal M 1996 *Pattern-Oriented Software Architecture: A System of Patterns*. John Wiley & Sons, Ltd.

Chakrapani LN, Korkmaz P, III VJM, Palem KV, Puttaswamy K and Wong WF 2001 The emerging crisis in embedded processors: what can a poor compiler do? *CASES'01*, pp. 176–181. Georgia, USA.

Chang LP and Kuo TW 2004 An efficient management scheme for large-scale flash memory storage systems. *The 2004 ACM Symposium on Applied Computing*, pp. 862–868. ACM.

Chaoui J, Cyr K, Giacalone JP, de Gregorio S, Masse Y, Muthusamy Y, Spits T, Budagavi M and Webb J 2002 OMAP: Enabling multimedia applications in third generation (3G) wireless terminals. Technical Report SWPA001, Texas Instruments.

Chinnici R 2002 Java APIs for XML based RPC Java Specification Request 101.

Clements P and Northrop L 2002 *Software Product Lines – Practices and Patterns*. Addison Wesley.

Coulouris G, Dollimore J and Kindberg T 2001 *Distributed Systems – Concepts and Design*. Addison-Wesley.

Edwards L, Barker R and EMCC Software Ltd 2004 *Developing Series 60 Applications. A Guide for Symbian OS C++ Developers*. Addison Wesley.

Ellis J and Young M 2003 J2ME Web Services 1.0 Sun Microsystems.

Fialli J and Vajjhala S 2006 Java architecture for XML binding Java Specification Request 222.

Filman RE, Elrad T, Clarke S and Akşit M 2005 *Aspect-Oriented Software Development*. Addison-Wesley.

Furber SB 2000 *ARM System-On-Chip Architecture*. Addison Wesley.

Gabriel RP 1989 Draft report on requirements for a common prototyping system. *SIGPLAN Not.* **24**(3), 93–165.

Gamma E, Helm R, Johnson R and Vlissides J 1995 *Design Patterns: Elements of Reusable Object-Oriented Software*. Addison Wesley.

Gannon J, McMullin P and Hamlet R 1981 Data abstraction, implementation, specification, and testing. *ACM Trans. Program. Lang. Syst.* **3**(3), 211–223.

Gehrmann C and Stahl P 2006 Mobile platform security. *Ericsson Review. The Telecommunications Technology Journal* (2), 59–70.

Graff MG and van Wyk KR 2003 *Secure Coding. Principles & Practices*. O'Reilly & Associates, Inc.

Harrison R 2003 *Symbian OS for Mobile Phones*. John Wiley & Sons, Ltd.

Hartikainen VM 2005 Java application and library memory consumption. Master of Science Thesis, Tampere University of Technology.

Hartikainen VM, Liimatainen PP and Mikkonen T 2006 On mobile Java memory consumption. *Euromicro Conference on Parallel and Realtime Systems*, pp. 333–339. IEEE Computer Society.

Heath C 2006 *Symbian OS Platform Security: Software Development Using the Symbian OS Security Architecture*. John Wiley & Sons, Ltd.

Jaaksi A 1995 Implementing interactive applications in C++. *Software Practice & Experience* **25**(3), 271–289.

Jayanti VBK and Hadley M 2001 SOAP with attachments API for Java. Java Specification Request 67.

Jipping MJ 2002 *Symbian OS Communications Programming*. John Wiley & Sons, Ltd.

Jones R 1999 *Garbage Collection. Algorithms for Automatic Dynamic Memory Management*. John Wiley & Sons, Ltd.

Kangas E and Kinnunen T 2005 Applying user-centered design to mobile application development. *Communications of the ACM* **48**(7), 55–59.

Koenig A and Moo BE 2000 *Accelerated C++ – Practical Programming by Example*. Addison Wesley.

Kortuem G, Scheider J, Preuitt D, Thompson TGC, Fickas S and Segall Z 2002 When peer-to-peer comes face-to-face: collaborative peer-to-peer computing in mobile ad-hoc networks. Technical report, IEEE.

Krasner GE and Pope ST 1988 A cookbook for using the model-view-controller user interface paradigm in Smalltalk-80. *Journal of Object-Oriented Programming* pp. 26–49.

Mallick M 2003 *Mobile and Wireless Design Essentials*. John Wiley & Sons Inc.

Maruyama H, Tamura K, Nakamura Y, Uramoto N, Neyama R, Murata M, Kosaka K, Clark A and Hada S 2002 *XML and Java. Developing Web Applications*. Addison Wesley.

McGovern J, Sims O, Jain A and Little M 2006 *Enterprise Service Oriented Architecture – Concepts, Challenges, Recommendations*. Springer.

Microsoft Corporation 2005 Web Service Dynamic Discovery (WS-Discovery). At URL *http://msdn.microsoft.com/library/en-us/dnglobspec/html/WS-Discovery.pdf*.

Miles R 2004 *AspectJ Cookbook*. O'Reilly & Associates.

Mullender S 1993 *Distributed Systems*. Addison-Wesley.

Myers GJ, Badgett T, Thomas TM and Sandler C 2004 *The Art of Software Testing*. John Wiley & Sons, Ltd.

Najmi F 2002 Java API for XML registries. Java Specification Request 93.

Noble J and Weir C 2001 *Small Memory Software. Patterns for Systems with Limited Memory*. Addison Wesley.

Norman DA 1998 *The Invisible Computer: Why Good Products Can Fail, the Personal Computer Is So Complex, and Information Appliances Are the Solution*. MIT Press.

O'Grady MJ and O'Hare GMP 2004 Just-in-time multimedia distribution in a mobile computing environment. *IEEE Multimedia* **11**(2), 62–74.

Oki B, Pfluegl M, Siegel A and Skeen D 1993 The information bus: an architecture for extensible distributed systems. *ACM Operating Systems Review* **27**, 58–68.

Parnas DL 1972 On the criteria to be used in decomposing systems into modules. *Communications of the ACM* **15**(2), 1053–1058.

Parnas DL, Clements PC and Weiss DM 1985 The modular structure of complex systems. *IEEE Transactions on Complex Systems* **11**(3), 259–266.

Pashtan A 2005 *Mobile Web Services*. Cambridge University Press.

Pasricha S, Luthra M, Mokapatra S, Dutt N and Venkatasubramanian N 2004 Dynamic backlight adaptation for low-power hand-held devices. *IEEE Design and Test of Computers* **21**(5), 398–405.

Pernici B 2006 *Mobile Information Systems – Infrastructure and Design for Adaptability and Flexibility*. Springer.

Pesonen J, Katara M and Mikkonen T 2006 Production-testing of embedded systems with aspects. *Hardware and Software Verification and Testing. First International Haifa Verification Conference*, pp. 90–102. Springer.

Riggs R, Taivalsaari A and VandenBrink M 2001 *Programming Wireless Devices with the Java 2 Platform, Micro Edition*. Addison-Wesley.

Sadjadi SM, McKinley PK and Kasten EP 2002 MetaSockets: Run-time support for adaptive communication services. Technical Report MSU-CSE-02-22, Department of Computer Science, Michigan State University.

Sales J 2005 *Symbian OS Internals: Real-time Kernel Programming*. John Wiley & Sons, Ltd.

Salmre I 2005 *Writing Mobile Code*. Addison-Wesley.

Satyanarayanan M 1997 Mobile computing: where's the tofu?. *ACM Sigmobile*.

Savikko VP 2000 *Building EPOC Applications*. VTT Technical Research Centre of Finland (in Finnish).

Schiller J 2003 *Mobile Communications* 2nd edn. Addison-Wesley.

Seng JS and Tullsen DM 2003 The effect of compiler optimizations on Pentium 4 power consumption *Seventh Workshop on Interaction between Compilers and Computer Architectures*. IEEE.

Shackman M 2005a Platform security – a technical overview. Symbian Ltd.

Shackman M 2005b Publish and subscribe. Symbian Ltd.

Shaw M and Garlan D 1996 *Software Architecture. Perspectives of an Emerging Discipline*. Prentice Hall.

Simpson M 1996 The point-to-point protocol (ppp). IETF RFC 1661.

Singh I, Brydon S, Murray G, Ramachandran V, Violleau T and Stearns B 2004 *Designing Web Services with the J2EE 1.4 Platform. JAX-RPS, SOAP, and XML Technologies*. Addison-Wesley.

Smith JE and Nair R 2005 *Virtual Machines – Versatile Platforms for Systems and Processes*. Elsevier.

Spinczyk O, Gal A and Schröder-Preikschat W 2002 AspectC++: An aspect-oriented extension to the C++ programming language. *Proceedings of the Fortieth International Conference on Tools Pacific: Objects for Internet, Mobile and Embedded Applications – Volume 10*, pp. 53–60. Australian Computer Society.

Spolsky J 2004 *Joel on Software: And on Diverse and Occasionally Related Matters That Will Prove of Interest to Software Developers, Designers, and Managers, and to Those Who, Whether by Good Fortune or Ill Luck, Work with Them in Some Capacity*. Apress.

Sridhar T 2003 *Designing Embedded Communications Software*. CMP Books.

Stitchbury J 2004 *Symbian OS Explained: Effective C++ Programming for Smartphones*. John Wiley & Sons, Inc.

Sun Microsystems 1997 Java remote method invocation specification.

Suomela R, Räsänen E, Koivisto A and Mattila J 2004 Open-source game development with the multi-user publishing environment (MUPE) application platform In *Proceedings of the 3rd International Conference on Entertainment Computing 2004* (ed. Rauterberg M), pp. 308–320. Springer.

Surakka K, Mikkonen T, Järvinen HM, Vuorela T and Vanhala J 2005 Towards compiler backend optimization for low energy consumption at instruction level In *Ninth Symposium on Programming Languages and Software Tools* (ed. Verne V and Meriste M). University of Tartu.

Suttor J and Walsh N 2004 Java API for XML processing. Java Specification Request 206.

Tanenbaum AS and van Steen M 2002 *Distributed Systems – Principles and Paradigms*. Prentice Hall.

Tasker M, Allin J, Dixon J, Forrest J, Heath M, Richardson T and Shackman M 2000 *Professional Symbian Programming – Mobile Solutions on the EPOC Platform*. Wrox Press.

Tiwari V, Malik S and Wolfe A 1994 Power analysis of embedded software: a first step towards software power minimization. *IEEE/ACM International Conference on Computer-Aided Design*, pp. 384–390. San Hose, California.

Topley K 2002 *J2ME in a Nutshell*. O'Reilly.

WAP Forum 2001 Wireless application protocol architecture specification WAP-210-WAPArch-20010712.

Whittaker J and Thompson H 2003 *How to Break Software Security*. Addison-Wesley.

Yoder J and Barcalow J 1997 Architectural patterns for enabling application security. *The 4th Pattern Languages of Programming Conference*. Monticelo, Illinois, USA.

Index

Printed and bound by CPI Group (UK) Ltd, Croydon, CR0 4YY

27/10/2024

14580148-0001